Consideration of Abiotic Natural Resources in Life Cycle Assessments

Consideration of Abiotic Natural Resources in Life Cycle Assessments

Special Issue Editor

Mario Schmidt

MDPI • Basel • Beijing • Wuhan • Barcelona • Belgrade

MDPI

Special Issue Editor
Mario Schmidt
Pforzheim University
Germany

Editorial Office
MDPI
St. Alban-Anlage 66
4052 Basel, Switzerland

This is a reprint of articles from the Special Issue published online in the open access journal *Resources* (ISSN 2079-9276) from 2015 to 2019 (available at: https://www.mdpi.com/journal/resources/special_issues/life-cycle)

For citation purposes, cite each article independently as indicated on the article page online and as indicated below:

LastName, A.A.; LastName, B.B.; LastName, C.C. Article Title. *Journal Name* **Year**, *Article Number,* Page Range.

ISBN 978-3-03897-545-8 (Pbk)
ISBN 978-3-03897-546-5 (PDF)

Cover image courtesy of Fotolia user Royce Bair.

Contents

About the Special Issue Editor

Mario Schmidt (Prof. Dr.) has been Professor at Pforzheim University (Germany) since 1999 and is the Director of the Institute of Industrial Ecology (INEC). He studied physics at the universities of Freiburg and Heidelberg. Since 1985, he has worked in several fields of environmental issues at IFEU Institute for Energy and Environmental Research in Heidelberg. In 1989 and 1990, he joined the Ministry of Environment of the State of Hamburg. Afterwards, he built up a research group on the Life Cycle Assessment at IFEU. From 2002 until 2014, he was responsible for research at Pforzheim University and spokesman of all 20 institutes for applied research (IAF) in the state of Baden-Wuerttemberg. In 2011, he founded the bachelor study program "Resource Efficiency Management", and in 2014 he built up the master study program "Life Cycle and Sustainability". Today he is leading a Ph.D. program "Energy Systems and Resource Efficiency", together with K.I.T. in Karlsruhe. In 2012, he was appointed to the Council on Sustainable Development of the federal state of Baden-Wuerttemberg. Since 2015, he has been also honorary professor at the Faculty of Sustainability of Leuphana University Lueneburg. He is member of various committees, e.g., of the German Association of Engineers (VDI) or ISO standardization committees.

Preface to "Consideration of Abiotic Natural Resources in Life Cycle Assessments"

On a global scale, there is a lot of talk about fake news at the moment. The perfidious thing is that even scientific theories with a high level of empirical evidence and methodically demanding proof of evidence are suddenly presented to the public as arbitrary. Climate change caused by greenhouse gases emitted by humans is one such example. However, scientific theories cannot be voted on democratically; nor can they be normatively defined or prohibited by a party programme or ideological opinions. They are and must remain part of an inner-scientific discourse.

Accordingly, science must keep its standards for evaluating theories and facts high, ensuring their quality and being transparent and comprehensible to the public. Scientific discourse is of great importance here, especially when theories are not yet clearly backed up by facts but are based on assumptions and plausibility. If the facts are not sufficient to provide clear evidence, there is great danger that one will be too strongly guided by ideological views or particular social interests.

In my opinion, one such area is that of resource depletion. Since the completion of the work of Dennis Meadows and his colleagues for the Club of Rome, the exhaustion of metal raw materials in the near future has become a very plausible assumption, with some factual support. However, there are also facts that contradict this. We cannot give a clear answer today. One could even cynically say that evidence can finally only be provided when the case has arisen—in other words, when natural resources have been exhausted. From this point of view, a cautious approach, i.e., the economical use of resources, would be the most sensible. However, that does not relieve science of the burden of dealing with the facts in a proper way. Actually, this is even more difficult because the resource sector is subject to strong economic interests, but at the same time it is also the focus of political and ideological positions.

This topic also has practical relevance. Many technologies that are promoted for reasons of climate protection have a poor resource balance. Within the Life Cycle Assessment, there can therefore be a trade-off between different environmental goals. In climate protection, mankind must react in short-term, i.e., within a few decades, and with drastic measures, i.e., a complete substitution of fossil fuels. However, how much time do we have on the question of resources? A scientific answer to this question requires discourse. Different opinions must be heard and discussed, in our case even from different disciplines. One's own assumptions must be questioned again and again. Are they still correct in view of the current facts?

This Special Issue of the journal Resources served this purpose and is fully documented in this book publication. An open and honest discussion also took place in London in 2015 at the Natural History Museum (see editorial). It has not yet come to an end and must be continued. I would like to thank everyone who has contributed to it so far and who is open to further discussion.

Mario Schmidt
Special Issue Editor

resources

MDPI

Editorial

Scarcity and Environmental Impact of Mineral Resources—An Old and Never-Ending Discussion

Mario Schmidt

Institute for Industrial Ecology, Pforzheim University, Tiefenbronner Str. 65, 75175 Pforzheim, Germany; mario.schmidt@hs-pforzheim.de; Tel.: +49-723-128-6406

Received: 6 December 2018; Accepted: 19 December 2018; Published: 21 December 2018

Abstract: A historical overview shows that mankind has feared the scarcity of mineral resources, especially metals, for many centuries. In the first half of the 20th century, this discussion was marked by the great military demand for raw materials, followed by the growing world population, increasing consumption and environmental awareness. From then on, there was less talk of regional shortages, but more discussion of a global scarcity or even a drying up of raw material sources worldwide. Although these forecasts are still controversially discussed today, the assessment of resource depletion has become an integral element of Life Cycle Assessments (LCA) or Life Cycle Impact Assessments (LCIA) of product systems. A number of methodological approaches are available for this purpose, which are presented and applied in a series of articles as part of a special issue of "Resources". The fundamental question is also addressed, namely to what extent the assessment of resource depletion in the context of an environmental study such as LCA is appropriate.

Keywords: resource depletion; critical materials; history; life cycle assessment

1. Introduction

In October 2015, a workshop entitled "Mineral Resources in LCIA: Mapping the path forward" took place in London. Richard Herrington of the Natural History Museum London and Johannes Drielsma of Euromines organized a meeting in which geologists, mining experts, and environmental scientists came together to present their different views on how to handle mineral resources, but also to look for commonalities. A fruitful discussion arose and the idea came up to record some of the thoughts. These records lead to a special issue of the magazine "Resources", which was published online in 2016. At the same time, there were other activities on this topic, e.g., various studies of the German Federal Environmental Agency, which contributed to the magazine with an article and delivered another paper in 2017 [1]. Meanwhile, there are other important publications on the subject, e.g., a statement and a joint report by the German Academy of Sciences of Technology, the National Academy of Sciences Leopoldina, and the Union of German Academies of Sciences [2,3], as well as a previously unreleased report by an international working group of experts of Life Cycle Assessment (LCA) [4]. There is still no uniform opinion on this important issue, but there is an intensive debate involving many different disciplines. The topic of scarcity and supply reliability has been discussed for a very long time, again and again, which proves a look into history. Therefore, in this introduction, not only are the contributions of the special edition briefly presented, but a reference to the long history of the discussion is made, and some rare sources are quoted by way of example.

2. Resource Scarcity in the Past

Two things make it so difficult to supply the industrial society with mineral raw materials and with metals in particular—they are limited on earth, and their extraction is associated with great effort and environmental pollution, both of which have concerned mankind for many centuries. In the

16th century, however, even a renewable resource threatened to become scarce—forests were cleared all over Europe because wood was the predominant raw material for mining and for the fires of melting furnaces. The Italian metallurgist Vannoccio Biringuccio (1480–1539) already warned in his "De la Pirotechnia" in 1540:

"I rather believe that someday people can no longer use the fire for the melting furnaces due to the lack of ores, because they process so much of it" (Chapter 10 in [5]).

It was not sure if the metal ores were a non-renewable resource at all. From the Italian island of Elba, which was an important iron ore deposit at the time, the following was said:

"With the quantities of ore that have been gained in so many centuries and still are gained, the mountains and islands would have to be completely leveled. Nevertheless, today more and better ore is produced than ever. Therefore, many believe that the ore, where it is mined, regenerates in the ground in a certain amount of time. If it is true, it would be something great, and it showed the great wisdom of nature and the great power of heaven". (Chapter 6 in [5]).

Unfortunately, it is not true, at least not in the time scale that is relevant to mankind. Thus, the search for the rare resource deposits has always been a great challenge to mankind. In this search, a variety of means has been employed, such as divining rods, which were already graphically demonstrated by the great German mining expert Georgius Agricola (1494–1555) (Figure 1). It was also Agricola who cited the critics of his time and described the environmental impact of mining and smelting:

"By mining for ore, the fields are devastated. By clearing the forests and groves, the birds and other animal species are eradicated. The ores are washed; but by this washing, because it poisoned the streams and rivers, the fish are either expelled from them or killed". (1st book in [6]).

Figure 1. Searching the lodes with the divining rod in the 16th century [6].

But Agricola, of course, defended mining, for it was already an important basis of civilization at that time. Previously, the use conflicts and the interventions in the landscape by mining were addressed by Paulus Niavis (1455–1517) [7].

The real scarcity of metal ores came during industrialization, when demand for metals sharply increased. Three hundred years ago, local shortages in England led to a nationwide ore trade [8].

The famous British economist Stanley Jevons posed the "coal question" in 1865. He saw limited coal supplies in face of rampant economic growth and advocated more moderate growth [9].

3. Critical and Strategic Metals

With modern times, the demand for metals increased immeasurably. In 1820, 1.65 million tons of pig iron was produced worldwide compared to 41 million tons in 1900, 250 million tons in 1960, and 1.2 billion tons today [10,11]. At the beginning of the twentieth century, a broad conservation movement was emerging in the U.S., focusing primarily on the limitations and protection of natural resources, including minerals, forests, soil, and fisheries, especially in the face of the rapidly growing US economy [12,13].

With the First World War, there was a growing concern in the United States that the supply of strategic raw materials could become difficult because international trade came partially to a halt [14]. An initial list of materials (Table 1), the supply of which could be of concern to the U.S., was published by C.K. Leith in 1917 for the War Industries Board [15]. The boundaries between military and industrial significance were still blurring. A second list of 42 materials was produced after World War I in 1921 by a committee led by General Harbord with a primarily military orientation [16]. The distinction between strategic and critical materials was first made in 1932. In 1939, the War Department compiled a list that included the term "essential material" [17]. The definitions were:

- **Strategic Materials** are those materials essential to the national defense for the supply of which in war dependence must be placed in whole, or in large part, on sources outside the continental limits of the United States, and for which strict conservation and distribution control measures will be necessary.
- **Critical Materials** are those materials essential in the national defense, the procurement problems of which in war, while difficult, are less serious than those of strategic materials, because they can be either domestically produced or obtained in more adequate quantities or have a lesser degree of essentiality, and for which some degree of conservation and distribution control will be necessary.
- **Essential Materials** are those materials essential to the national defense for which no procurement problems in war are anticipated, but whose status is such as to require constant surveillance because future developments may necessitate reclassification as strategic or critical.

Table 1. The first lists of strategic and critical materials in the U.S.

Leith List 1917
Deficient in a major degree: Tin, Nickel, Platinum and metals of the platinum group, Antimony, Vanadium, Zirconium, Mica, Monazite, Graphite, Asbestos, Ball clay and kaolin, Chalk, Cobalt, Naxos emery, Grinding pebbles. **Deficient in a lesser degree:** Nitrates (except potassium nitrates), Potash, Manganese, Chromite, Magnesite.
Harbord List 1921
Agar, Antimony, Arsenic, Asphalt, Balsa, Camphor, Chromium, Coconut Shells, Coffee, Cork, Graphite, Hemp, Hides, Iodine, Jute, Kapok, Linseed Oil, Manila Fiber, Mercury, Mica, Nickel, Nitrogen, Nux Vomica, Opium, Palm Oil, Phosphorus, Platinum, Potassium Nitrate, Quinine, Rubber, Silk, Manganese (Ferro-grade), Shellac, Sodium Nitrate, Sugar, Sulphur, Thymol, Tin, Tungsten, Uranium, Vanadium, Wool.
War Department List 1939
Strategic: Aluminum, Antimony, Chromium, Manganese, Mercury, Mica, Nickel, Tin, Tungsten. **Critical:** Asbestos, Cadmium, Cryolite, Fluorspar, Graphite, Iodine, Platinum, Titanium, Vanadium. **Essential:** Abrasives, Arsenic, Chlorine, Copper, Helium, Iron and Steel, Lead, Magnesium, Molybdenum, Ammonia and Nitric Acid, Petroleum, Phosphates, Potash, Refractories, Sulfur and Pyrite, Uranium, Zinc, Zirconium.

In the 1930s, several of the U.S.-governmental institutions' other authors recommended the creation of strategic stocks of so-called scarce minerals [17–19]. In 1939, the first federal law authorizing stockpiling of strategic materials was enacted in the U.S. This stockpiling exists still today in the U.S. and is operated by the National Defense Stockpile (NDS). The total inventory of the NDS represented a market value of $1.15 billion in 2016 [20].

Thus, the concept of critical materials was introduced, as well as the academic attention to the scarcity of industrial or defense-related raw materials. It was always more about the topic of which raw materials were available for the U.S. economy (or military forces) and less about how many raw materials were available worldwide.

The scarcity and availability of resources was then repeatedly addressed, e.g., with the "Road of Depletion", which was presented in a hearing of the U.S. Senate 1949 by James Boyd, the director of the U.S. Department of Mining (Figure 2) [21]. At that time, it was already very clear that only 7% of the world's population, namely in the U.S., use 50% of the world's minerals and 70% of the world's oil. The U.S. president installed a Materials Policy Commission, which in 1952 submitted a major report titled "Resources for Freedom" [22]. The Cold War was also a contest for economic power and access to natural resources. In 1963, a large systematic empirical study by Barnett and Morse of historic trends for various natural resources between 1870 and 1958 eventually supported the hypothesis of a decreasing (rather than an increasing) scarcity [23]. They represented a critical but nevertheless optimistic picture of the resource question. They believed in technical progress and in raising efficiency.

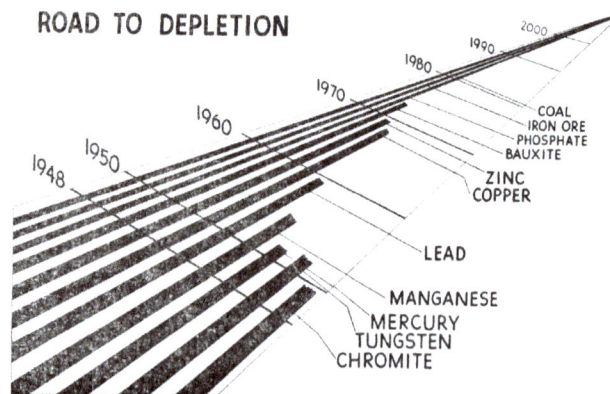

Figure 2. James Boyd: "The chart indicates the number of years of normal requirements our present known reserves of critical materials will supply" [21].

This optimistic picture changed fundamentally in the 60s through wake-up calls such as Ehrlich's book, "The Population Bomb" [24], but especially with the Club of Rome study by Meadows, "The Limits to Growth" in 1972 [25]. Limited natural resources would be confronted with an almost rampant growth of world population and global economic output. Now, it was increasingly about the global development, and the careless handling of the resources was criticized. For example, a study by the U.S. National Academy of Sciences (NAS) asked for increased recycling in 1969: "The automobile is a prime target for improvement. The copper content of the average car should be reduced from about 1.4 percent to 0.4 percent or less of the total carcass and problems of metal recovery simplified" [26]. Recycling became a guiding theme of environmental policy in the following decades.

In 1975, the NAS prepared another report on "Mineral Resources and the Environment" [27]. Not only was the scarcity of raw materials—both energetic and non-energetic—addressed, but also the environmental impact in particular, which was demonstrated by the example of coal extraction and use. A "conservation ethic" was demanded, which could just as well have been formulated today:

"Because of limits to natural resources as well as to means for alleviating these limits it is recommended that the Federal Government proclaim and deliberately pursue a national policy of conservation of material, energy and environmental resources, informing the public and the private sectors fully about needs and techniques for reducing energy consumption, the development of substitute materials, increasing the durability and maintainability of products, and reclamation and recycling" (page 37 in [27]).

The NAS pointed out that the stockpiling of materials in the past was mainly for military reasons. It was stated that "similar considerations can often be applied to the protection of the U.S. economy and the essential needs of the civilian sector" (page 34 in [27]). This had changed little until today.

The two updates to the Barnett & Morse study, "Scarcity and Growth Reconsidered" [28] and the study of Menzie, Singer, and DeYoung, Jr. in "Scarcity and Growth Revisited" [29] essentially confirmed the old results that there is no geological scarcity. Menzie et al. noted that the physical availability of resources in itself does not constitute a growth limit. However, the effort required to obtain them is growing, although many resources remain abound. It is obvious that supplies of mineral resources were first used most intensively in the areas closest to their use. As demand increased, exploration and eventually extraction across oceans in inhospitable climates, always deeper into land and water, occurred. Thus, costs, energy input, and the destruction of the environment associated with the extraction increased. Menzie et al. directed the attention to the fact that it is not the limited quantities of raw materials but the accompanying circumstances of their extraction that are the real problem.

Nevertheless, the image of the ebbing raw material sources became apparent to the public. The study by Meadows, which has made popular the very descriptive concept of resource lifetime [25], has contributed significantly to this. The Meadows team introduced the "static reserve life index", which states how many years the known reserves of a given resource will last when the current annual consumption is assumed. With the exponential index, a continuously increasing consumption is expected, which again significantly reduces the time the reserves are available. It hit a nerve with the public and was quoted from time to time, but it was also discussed controversially. Yet, Gerling and Wellmer found out that raw material lifetimes did not decrease over the decades, but mostly stayed the same or even increased [30]. The indicator describes the economic effects of exploration in the mining industry rather than a geological scarcity.

The discussion of the past decades was also marked by reports from the U.S. For its first report in 1988, the U.S. National Critical Materials Council, founded by president Ronald Reagan, selected seven key commodities from three basic categories into which strategic and critical materials were broadly divided [31]. These included: (1) critical alloys—cobalt, chromium, and ferrosilicon; (2) potential high growth security materials—germanium and titanium; and (3) high-volume materials—aluminum and copper. Again, the strategic importance of supply and demand, the current status of the so-called National Defense Stockpile, and the global situation of the import dependence and vulnerability of the U.S. economy were discussed.

Recently, various incidents have come together; countries in Asia, South America, and Africa are claiming an ever-increasing share of resources to build their economies and to supply their populations. The global commodity prices rose rapidly in the first decade of the 21st century, causing a public "resource shock". At the same time, technical innovation has become increasingly dependent on the quantity and variety of raw materials. Many high-tech products have become indispensable in today's life, but they cannot be produced without certain raw materials. A possible scarcity of raw materials endangers not only the military-strategic position of nations, but the way-of-life of the previously wealthy countries and their primacy in the technological development of new products. In addition, there is a global ecological conscience that questions the social and ecological consequences of the use of resources.

These thoughts have been reflected in the report "Minerals, Critical Minerals, and the U.S. Economy" of the Committee on Critical Mineral Impacts of the U.S. Economy of the National Research Council, which was published in 2008 [32]. The report developed the current method of the semi-qualitative description of the criticality of raw materials with a multi-dimensional evaluation matrix. The impact of supply restriction is plotted against the supply risk as a two-dimensional graph and determined individually for the various raw materials.

An important boost provided the work of the International Resource Panel of United Nations Environmental Programme (UNEP), which published several reports on the subject of metal resources between 2009 and 2014 [33] and in particular called for increased efforts to recycle. In 2016, the U.S. National Science and Technology Council (NSTC) published a report that provided a systematic methodology for screening potentially critical minerals [34,35]. Another detailed report was issued by the U.S. Geological Survey in 2017 [36]. In Europe, a corresponding list was issued by the European Commission. The first list of "Critical Raw Materials" was prepared in 2010; updates were made in 2014 and 2017 [37–39]. Most recently, the 2017 assessment included a total of 78 individual materials.

The disadvantage of these presentations is that they are only short to medium term aligned, and thus the long-term supply situation is not taken into account for the obvious and above mentioned reasons. Furthermore, it is purely economically oriented and ecological aspects are largely missing.

It was Thomas Graedel and his team who developed a three-dimensional criticality system in which the environmental impact has its own dimension [40]. A similar approach was recently published by the German Federal Environment Agency [41]. It is currently being applied to a variety of chemical elements. Results can be expected for 2019. However, it is already evident that the environmental impact associated with the extraction and processing of raw materials can hardly be described with cardinal scales, as is known from LCA. For this purpose, too many site-specific qualitative aspects, at the mining sites for example, have to be considered. This makes the implementation in the framework of LCA difficult.

What can we learn from the past? The scarcity of resources is not new. Concern for the drying up of raw materials is probably as old as mankind itself. The striking of raw materials has always been associated with labor and effort. The estimation of scarcity in each epoch was always done against the background of the respective knowledge available, but the interests connected with the raw materials were also very decisive. It becomes very clear that, especially in the past 100 years, the military interests played an important role and still do today. Many high-tech products that require special raw materials are indispensable to the military. They have a strategic meaning. In the public and scientific discussion, however, it is argued as increasingly "civil" and linked to the material and energy-intensive "way-of-life". An important role is played by the LCA of products and services, which quantifies the impact on the environment. The use of resources is an integral part of the analysis and evaluation.

4. Abiotic Resources in Life Cycle Assessment (LCA)

When an LCA is carried out for products or services, it is now standard practice to include and quantify the use of natural resources. The Life Cycle Inventory still does this on a physical and quantitative basis, i.e., the amount of required raw materials and the withdrawals from nature is quantified. For example, the use of water as a natural resource is included. It has also become customary to consider the required land use. However, the pure quantities (m^3 of water or km^2 of area) are not sufficient to describe the environmental quality of the resource input, yet this is needed in the following step of an LCA, the Life Cycle Impact Assessment (LCIA), where the ecological relevance of the energy and material flows is quantified. To get from the amount of a substance to the effect of the substance, so-called characterization factors are used in the LCA. They are a simplification for the LCA calculation, and all the knowledge about the ecological effect of a substance is hidden behind them. Their investigation is therefore always in the focus of the interests of many authors from the LCA community.

This task also arises for the mineral resources taken from the lithosphere. The input of metals that originate from nature and enter the technosphere is one thing, but what is the ecological relevance of the volume flows of iron, copper, tantalum, indium, gold, etc.? The energy demands, the wastewater, and the emissions associated with the extraction and processing of raw materials are already included in an LCA. These environmental aspects of mining and metalworking are automatically considered; thus double counting must be avoided. Rather, it is about the question of how the extraction of raw materials from the lithosphere "in itself" can be evaluated.

In the field of Life Cycle Assessments, the safeguard objects and the so called "Areas of Protection" (AoP) have been discussed for many years [42–44]. It is not only interesting to know what impact a human action has on the climate, the acid rain, or the eutrophication, but what that impact means for the safeguard objects, especially for human health and for the integrity of nature, which is often circumscribed with the preservation of biodiversity. In addition, there is a third safeguard object, namely the preservation of natural resources [45–47]. Strictly speaking, this is not an ecological aspect, but it is more subject to the idea of sustainability. The consumption of a limited natural resource eventually leads to its depletion. What is not kept in the cycle of nature disappears at some point and is no longer available for future generations, which would not meet the idea of sustainability.

However, does the mining of minerals and possibly the depletion of metals really belong to an ecological analysis like the Life Cycle Assessment? Are these not rather socio-economic aspects that cannot be adequately illustrated with the methodological instruments of the LCA? This issue has been the subject of much controversy for many years. In an attempt to hierarchize the safeguard objects in the life cycle assessment, Hofstetter and Scheringer (1997) based the LCA on human welfare and divided it into the social welfare of today's generations and the material welfare of future generations [48]. They identified additional safeguard objects related to resource supplies, human health, biodiversity, and ecological health (Figure 3).

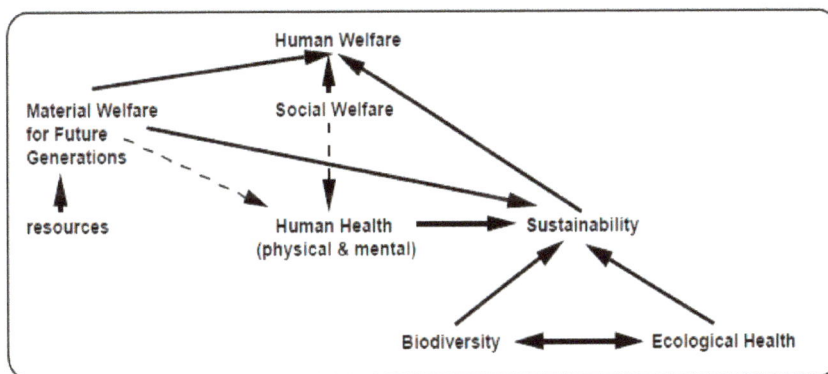

Figure 3. Proposal for the hierarchy of safeguard objects in the Life Cycle Assessment according to [48].

The safeguard object "resources" could be interpreted in such a way that a reduction of the resource supply or the lower quality of future mineral deposits restricts the freedom of action of coming generations and imposes higher efforts on them. Especially with regard to sustainability and intergenerational justice, the protection of mineral resources would then be worth considering. It is controversial if this should be considered in an environmental analysis such as the LCA. Social issues, for example, are excluded from the LCA and treated with their own instrument, the "Social LCA". If there were a suitable instrument to assess economic sustainability, such as influencing the welfare of future generations through current activities, the resources would have to be considered in this assessment. However, this instrument does not exist. This can be seen as a justification for today's explicit inclusion of resources as safeguard objects by the LCA methodology as a stopgap for the inability of economics, so to speak.

The basic aspect of whether or not resources should be considered as safeguard objects is hardly questioned by the LCA community today [4]. One may criticize this because it is based on the claim to model nearly everything with the LCA that concerns the metabolism of the technosphere and its exchange with the biosphere. There are far more relevant questions to be asked, such as whether the simple linear LCA approach can adequately address the dynamic effects of technological innovation or market developments. Another difficulty is finding suitable indicators for a quantitative assessment within the LCA. Finally, the amount of remaining inventory or usage restriction for future generations would have to be quantified in some way. However, this is largely unknown today, just as in the past (see Chapter 1) it was unknown what resources would be available to us in the 20th or 21st century.

In summary, three questions play a role at the interface between the resource topic and the Life Cycle Assessment:

1. How scarce are the mineral resources, and in particular the metals, and do we really know the amount of mineral resources left in the earth?
2. Do we have to understand the mineral resources in nature as protective goods that, in addition to health and biodiversity, have to be protected and preserved for reasons of sustainability?
3. Which environmental impacts occur through the mining of resources and the extraction of metals, and are they adequately reflected in the method of the Life Cycle Assessment?

5. Contributions to the Main Topic

The discussion at the 2015 workshop in London triggered this discussion and is well documented by some articles from participants. There are also additional contributions that round off the topic altogether. Drielsma et al. [49] gave an overview of the discussion in London that included the points of contact of the various scientific disciplines, the different perspectives on the subject, and the difficulties of definition that sometimes complicate scientific exchange.

Meinert et al. [50] advanced a point of view that can be found among many geologists. They described how mineral resources are explored and discovered, which leads to predictions about known deposits and which definitions are used for them. They used the example of copper because very different opinions about the scarcity of this metal exist and it has been mined in large quantities for a long time. In this example, they tried to show that the lifetime concept—or the peak concept—is based on wrong assumptions and therefore leads to misinterpretations. According to their statements, by 2050, only half of the previously known and already economically degradable stocks will be needed, and the undiscovered copper deposits are not yet included. By estimating these deposits, their optimistic prognosis was that primary copper will still be available for many generations. The authors argued that less concern should be paid to the depletion of primary resources but rather what happens to resources after their extraction and how they are used with regard to dissipation. On this point, there is certainly a broad consensus with other experts who have a more "resource-pessimistic" attitude. But the most important point of their contribution may be to suggest that society is investing too little in education, research, and development to ensure the supply of raw materials for future generations. The mining sector in particular needs stronger state support. When more environmentally compatible and efficient mining methods are used, there will certainly be broad consensus on this point.

The article by Oers and Guinée [51] was particularly special because it was a kind of update and reflection on an approach that is widely used in the LCA community. In 1995, Guinée and Heijungs proposed characterization factors for abiotic depletion potential (ADP), which were widely used in the application of the LCA [52]. Again, the use of terms played an important role. Do we talk about depletion, scarcity, or criticality? Should the ultimate reserves, the reserve base, or the crustal content be used as a basis for comparisons between different metals? The authors emphasized that there is probably more than a dilution problem, namely when resources are released into the environment through emissions or wastes and are irretrievably lost. However, they also pointed out that it is difficult to define the correct method because the parameters to be chosen depend on the question and cannot be empirically verified in practice.

Calvo et al. [53] drew attention to an interesting aspect—what does the mineral capital of various countries look like? Through the mining and export of minerals, this capital is changing. What exergy would be needed to rebuild this lost mineral capital? An average concentration of minerals in the earth's crust is assumed. The authors created a kind of a mineral balance based on exergy replication costs. Evaluating the exergy with prices (for electricity or coal) leads to an economic statement. Using the examples of Colombia and Spain, the authors showed that mining and exporting minerals produces a lower gross domestic product than one would have to pay for exergy replacement costs to rebuild the mineral capital. The figures could also be used to produce net balances between countries. This issue has a high economic and developmental importance due to the unequal distribution of raw materials among countries and the question of fair pay for raw materials supplies.

Vieira et al. [54] calculated the surplus costs arising from current resource extraction on future situations and used these as a basis to calculate characterization factors for 12 metals and the platinum group metals. In the tradition of many environmental scientists, they assumed the absolute finiteness of mineral resources. They further assumed that with increasing mine production, the ore grades decrease and a grade-tonnage relationship can be set up. They derived a function for the operating costs per metal extracted, which depends on the previous amount of cumulative metal extracted and the total amount of a metal that can be mined on earth or has already been mined. The interesting thing about the approach is that the choice of this last value has only a limited effect on the result. The authors pointed out that they had not yet considered many cost drivers and therefore need to gather more data.

At the focus of the article by Henßler et al. [55] was the application of a method called ESSENZ, which was developed at the Technical University of Berlin to evaluate many aspects of the use of resources. These aspects included physical availability as well as socioeconomic availability and environmental impacts. A total of 18 categories were taken into account, some category examples being abiotic resource depletion, the political stability of producing countries, and the impact of summer smog. The authors presented a case study in which ESSENZ was applied to the comparison of a conventional car and an e-car from Mercedes. The method provides greatly differentiated results for the different categories. In particular, it allows the comparison of tradeoffs that may occur when environmental impacts are reduced, but the use of resources increases at the same time.

In their article, Martin-Gamboa and Iribarren [56] examined and compared the performance of wind turbines, taking into account the use of raw materials. The starting point was, of course, data from the LCA, but they used a method that goes far beyond the pure LCA approach. They used emergy as an indicator, which is the solar energy that is or was ultimately required to manufacture a product by extracting resources from the geo-biosphere. This approach is similar to corresponding exergy approaches in resource use assessment. This is connected, as they wrote, with a departure from the purely anthropocentric perspective. Gamboa and Iribarren went even further and used the data envelopment analysis (DEA) for time-dependent efficiency measurements of various wind farms.

Alvarenga et al. [57] analyzed, in a very extensive study, the different methods with which abiotic raw materials are evaluated in LCA. They also dealt with the question of the area of protection and which impact assessments are useful. In total, they considered 19 different Life Cycle Impact Assessment methods, some of which treat the topic of resources very differently. They tested the methods using a case study comparing fossil and bioethanol-based ethylene production.

Finnveden et al. [58] made a very important contribution by considering the crucial question of what is actually depleted or consumed. Their answer was that it is neither matter nor energy, but usable energy, i.e., exergy. The big advantage to this is that the evaluation of matter and energy can be integrated into a unified concept. However, this thermodynamic approach requires extensive calculations. The question that remains open is what practical relevance the calculation of the more theoretical exergy values has. This must be further discussed in the future at case studies.

Müller et al. [1], in a very committed, detailed, and far-reaching article, dealt with the fundamental questions of which goals one wants to achieve with dematerialization and which indicators are meaningful for this purpose. They dealt with normative aspects in the discussion, cited examples from German and European resource efficiency policies, and came to the conclusion that mass-based indicators are rather unsuitable to describe the desired AoP of environmental protection. At the same time, however, they also advocated an environmental policy that does not stop at national borders but takes a life cycle perspective.

6. Outlook

The presented articles show that the question of the depletion of resources and their assessment in the LCA is by no means conclusively answered. There are many opinions and different approaches. Above all, however, it is an interdisciplinary question that has a lot to do with methods and experiences from environmental sciences, but most significantly with geology, resource economics, and mining. The positions on the depletion question depend very much on different specialists' knowledge, but also on the normative positions of the various disciplines. It is therefore quite understandable that representatives from the environmental sciences assume that resources will soon be exhausted. Representatives from the mining sector, on the other hand, must start from the inexhaustibility of resources. This is the basis of their business model.

Nevertheless, recommendations for politics and for the economy are necessary, such as which criteria are used to assess recycling strategies or a circular economy. If the depletion of resources is an independent and urgent issue, a circular economy would have absolute priority. In this case, this might have to be weighed against other objectives, e.g., climate protection. Conflicting goals could arise here since the achievement of the 1.5° goal requires an enormous expansion and the restructuring of energy supply structures worldwide. This would require a considerable use of resources that could only be covered by additional primary resources. If, however, the problem of resource depletion is postponed or perceived only as an exergy or entropy problem, then the energy balance and the environmental impacts of resource supply come to the fore, both from mining and from recycling. These can be dealt with very well in today's LCA framework.

This is why the discussion on this topic is more important than ever. However, this time it must lead to a result and must not get stuck in different schools of thought as it did 40 years ago. Above all, the exchanges between the various scientific disciplines and applications in economy are important. Therefore, the meeting in London, which resulted in most of the articles presented in this paper, was an important event that urgently needs to be repeated on an interdisciplinary and open platform.

Funding: This research received no external funding.

Conflicts of Interest: The author declares no conflict of interest.

References

1. Müller, F.; Kosmol, J.; Keßler, H.; Angrick, M.; Rechenberger, B. Dematerialization—A disputable strategy for resource conservation put under scrutiny. *Resources* **2017**, *6*, 68. [CrossRef]
2. German National Academy of Sciences Leopoldina; National Academy of Science and Engineering (acatech); Union of the German Academies of Sciences and Humanities. *Raw Materials for the Energy Transition*; Acatech: Munich, Germany, 2018.
3. Wellmer, F.-W.; Buchholz, P.; Gutzmer, J.; Hagelüken, C.; Herzig, P.; Littke, R.; Thauer, R.K. *Raw Materials for Future Energy Supply*; Springer: Berlin, Germany, 2018.
4. Berger, M.; Sonderegger, T. *Global Guidance for Life Cycle Impact Assessment of Mineral Resource Use. White Paper of the Task Force Resources of Life Cycle Initiative*; Berger, M., Sonderegger, T., Eds.; Life Cycle Initiative of UN Environment: Paris, France, 2018; unpublished.
5. Johannsen, O. *Biringuccios Pirotechnia*; Vieweg: Braunschweig, Germany, 1925. (In German)

6. Agricola, G. *De Re Metallica Libri XII, Basel, 1556*, 2nd German ed.; Ludwig Königs: Basel, Switzerland, 1621. (In German)

7. Niavis, P. Iudicium Iovis. In *Paulus Niavis Iudicium Iovis oder Das Gericht der Götter über den Bergbau Freiberger Forschungshefte D3*; Krenkel, P., Ed.; Akademie-Verlag: Berlin, Germany, 1953. (In German)

8. Selmeier, F. *Eisen, Kohle und Dampf. Die Schrittmacher der Industriellen Revolution*; Deutsches Museum: München, Germany, 1984; p. 87. (In German)

9. Jevons, W.S. *The Coal Question*, 3rd ed.; Sentry Press: New York, NY, USA, 1965.

10. Riegraf, H. *Geographie der Rohstoffwirtschaft der Erde*; Deutscher Verlag der Wissenschaften: Berlin, Germany, 1961; p. 282. (In German)

11. World Steel Association. *Steel Statistical Yearbook 2018*; World Steel Association: Brussels, Belgium, 2018.

12. Pinchot, G. *The Fight for Conservation, Doubleday*; Page & Company: New York, NY, USA, 1911.

13. Blanchard, N.C. (Ed.) *Proceedings of a Conference of Governors in the White House Washington, DC 13–15 May 1908*; Government Printing Office GPO: Washington, DC, USA, 1909.

14. Leith, C.K. *The Economic Aspects of Geology*; Henry Holt and Company: New York, NY, USA, 1921.

15. Leith, C.K. International Control of Minerals. In *Mineral Resources of the United States*; McCaskey, H.D., Ed.; Department of the Interior: Washington, DC, USA, 1917; pp. 7–17.

16. Spicer, L.G. Stock Piling of Strategic Materials. Master's Thesis, Boston University, Boston, MA, USA, 1950.

17. Roush, G.A. *Strategic Mineral Supplies*; McGraw-Hill: New York, NY, USA, 1939.

18. Emeny, B. *The Strategy of Raw Materials*; MacMillan: New York, NY, USA, 1936.

19. Eckes, A.E. *The United States and the Global Struggle for Minerals*; University of Texas Press: Austin, TX, USA, 1979.

20. U.S. Department of Defense, DLA Strategic Materials. *Strategic and Critical Materials Operations Report to Congress*; Fort Belvoir: Fairfax, VA, USA, 2017.

21. U.S. Senate. *Mineral Resources Development. Hearings before the Committee on Interior and Insular Affairs*; U.S. Senate: Washington, DC, USA, 1949.

22. House of Representatives. *Resources for Freedom*; A Report to the President by The President's Materials Policy Commission; House of Representatives: Washington, DC, USA, 1952; Volume I–V.

23. Barnett, H.J.; Morse, C. Scarcity and Growth. In *The Economics of Natural Resource Availability*; John Hopkins Press: Baltimore, Maryland, 1963.

24. Ehrlich, P.R. *The Population Bomb*; Sierra Club: San Fransisco, CA, USA, 1968.

25. Meadows, D.H.; Meadows, D.L.; Randers, J.; Behrens, W.W. *The Limits to Growth*; Universe: New York, NY, USA, 1972.

26. National Academy of Sciences. *Resources and Man*; Freeman: San Francisco, CA, USA, 1969.

27. National Academy of Sciences. *Mineral Resources and the Environment*; A Report by the Committee on Mineral Resources and the Environment; National Academy of Sciences: Washington, DC, USA, 1975.

28. Smith, V.K. *Scarcity and Growth Reconsidered*; John Hopkins Press: Baltimore, Maryland, 1979.

29. Simpson, R.D.; Toman, M.A.; Ayres, R.U. Scarcity and Growth Revisited. In *Natural Resources and the Environment in the New Millennium*; Resources for the Future: Washington, DC, USA, 2005.

30. Gerling, J.P.; Wellmer, F.-W. Reserven, Ressourcen und Reichweiten. Wie lange gibt es noch Erdöl und Erdgas? *Chem. Uns. Zeit.* **2005**, *39*, 236–245. (In German) [CrossRef]

31. Executive Office of the President National Critical Materials Council. *A Critical Materials Report—The Continuation of a Presidential Commitment*; Executive Office of the President National Critical Materials Council: Washington, DC, USA, 1988.

32. Committee on Critical Mineral Impacts of the U.S. Economy. *Minerals, Critical Minerals, and the U.S. Economy*; National Research Council: Washington, DC, USA, 2008.

33. International Resource Panel. *Work on Global Metal Flows*; United Nations Environment Programme: Paris, France, 2013.

34. Committee on Environment, Natural Resources, and Sustainability. *Assessment of Critical Minerals: Screening Methodology and Initial Application*; National Science and Technology Council: Washington, DC, USA, 2016.

35. Committee on Environment, Natural Resources, and Sustainability. *Assessment of Critical Minerals: Updated Application of Screening Methodology*; National Science and Technology Council: Washington, DC, USA, 2018.

36. U.S. Geological Survey. *Critical Mineral Resources of the United States—Economic and Environmental Geology and Prospects for Future Supply*; USGS: Reston, VA, USA, 2017.

37. European Commission. *Tackling the Challenges in Commodity Markets and on Raw Materials*; COM(2011) 25 final; European Commission: Brussels, Belgium, 2011.

38. European Commission. *On the Review of the List of Critical Raw Materials for the EU and the Implementation of the Raw Materials Initiative*; COM(2014) 297 final; European Commission: Brussels, Belgium, 2014.

39. European Commission. *On the 2017 List of Critical Raw Materials for the EU*; COM(2017) 490 final; European Commission: Brussels, Belgium, 2017.

40. Graedel, T.E.; Barr, R.; Chandler, C.; Chase, T.; Choi, J.; Christoffersen, L.; Friedlander, E.; Henly, C.; Jun, C.; Nassar, N.T.; et al. Methodology of Metal Criticality Determination. *Environ. Sci. Technol.* **2012**, *46*, 1063–1070. [CrossRef] [PubMed]

41. Manhart, A.; Vogt, R.; Priester, M.; Dehoust, G.; Auberger, A.; Blepp, M.; Dolega, P.; Kämper, C.; Giegrich, J.; Schmidt, G.; et al. The environmental criticality of primary raw materials—A new methodology to assess global environmental hazard potentials of minerals and metals from mining. *Min. Econ.* **2018**. [CrossRef]

42. Consoli, F. *Guidelines for Life Cycle Assessment: A "Code of Practice"*; SETAC (Society of Environmental Toxicology and Chemistry): Pensacola, FL, USA, 1993.

43. Beltrani, G. Safeguard Subjects. The Conflict Between Operationalization and Ethical Justification. *Int. J. Life Cycle Assess.* **1997**, *2*, 45–51. [CrossRef]

44. Vadenbo, C.; Rørbech, J.; Haupt, M.; Frischknecht, R. Abiotic resources: New impact assessment approaches in view of resource efficiency and resource criticality—55th Discussion Forum on Life Cycle Assessment, Zurich, Switzerland, 11 April 2014. *Int. J. Life Cycle Assess.* **2014**, *19*, 1686–1692. [CrossRef]

45. Steen, B.A. Abiotic resource depletion. Different perceptions of the problem with mineral deposits. *Int. J. Life Cycle Assess.* **2006**, *11*, 49–54. [CrossRef]

46. Strauss, K.; Brent, A.C.; Hietkamp, S. Characterisation and normalisation factors for life cycle impact assessment mined abiotic resources categories in South Africa. *Int. J. Life Cycle Assess.* **2006**, *11*, 162–171. [CrossRef]

47. Mancini, L.; De Camillis, C.; Pennington, D. Security of supply and scarcity of raw materials. In *Towards a Methodological Framework for Sustainability Assessment*; European Commission Joint Research Centre: Ispra, Italy, 2013.

48. Hofstetter, P.; Scheringer, M. *Schutzgüter und ihre Abwägung aus der Sicht verschiedener Disziplinen*; 5. Diskussionsforum Ökobilanzen vom 17. Oktober 1997 an der ETH Zürich; ETH Zürich: Zürich, Switzerland, 1997. (In German)

49. Drielsma, J.A.; Allington, R.; Brady, T.; Guinée, J.; Hammarstrom, J.; Hummen, T.; Russell-Vaccari, A.; Schneider, L.; Sonnemann, G.; Weihed, P. Abiotic raw-materials in life cycle impact assessments: An emerging consensus across disciplines. *Resources* **2016**, *5*, 12. [CrossRef]

50. Meinert, L.D.; Robinson, G.R., Jr.; Nassar, N.T. Mineral resources: Reserves, peak production and the future. *Resources* **2016**, *5*, 14. [CrossRef]

51. Van Oers, L.; Guinée, J. The abiotic depletion potential: Background, updates, and future. *Resources* **2016**, *5*, 16. [CrossRef]

52. Guinée, J.; Heijungs, R. A proposal for the definition of resource equivalency factors for use in product Life-Cycle Assessment. *Environ. Toxicol. Chem.* **1995**, *14*, 917–925. [CrossRef]

53. Calvo, G.; Valero, A.; Carmona, L.G.; Whiting, K. Physical assessment of the mineral capital of a nation: The case of an importing and an exporting country. *Resources* **2015**, *4*, 857–870. [CrossRef]

54. Vieira, M.D.M.; Ponsioen, T.C.; Goedkoop, M.J.; Huijbregts, M.A.J. Surplus cost potential as a life cycle impact indicator for metal extraction. *Resources* **2016**, *5*, 2. [CrossRef]

55. Henßler, M.; Bach, V.; Berger, M.; Finkbeiner, M.; Ruhland, K. Resource efficiency assessment—Comparing a plug-in hybrid with a conventional combustion engine. *Resources* **2016**, *5*, 5. [CrossRef]

56. Martin-Gamboa, M.; Iribarren, D. Dynamic ecocentric assessment combining emergy and data envelopment analysis: Application to wind farms. *Resources* **2016**, *5*, 8. [CrossRef]

57. Alvarenga, R.A.F.; de Olivera Lins, I.; de Almeida Neto, J.A. Evaluation of abiotic resource LCIA methods. *Resources* **2016**, *5*, 13. [CrossRef]
58. Finnveden, G.; Arushanyan, Y.; Brandao, M. Exergy as a measure of resource use in life cycle assessment and other sustainability assessment tools. *Resources* **2016**, *5*, 23. [CrossRef]

resources

MDPI

Communication

Abiotic Raw-Materials in Life Cycle Impact Assessments: An Emerging Consensus across Disciplines

Johannes A. Drielsma [1,*], Ruth Allington [2,†], Thomas Brady [3,†], Jeroen Guinée [4,†],
Jane Hammarstrom [5,†], Torsten Hummen [6,†], Andrea Russell-Vaccari [7,†],
Laura Schneider [8,†], Guido Sonnemann [9,†] and Pär Weihed [10,†]

1   European Association of Mining Industries, Metal Ores and Industrial Minerals (Euromines),
    Avenue de Broqueville/Broquevillelaan 12, Brussels 1150, Belgium
2   Committee for Mineral Reserves International Reporting Standards (CRIRSCO), c/o Pan-European Reserves
    & Resources Reporting Committee (PERC), c/o EFG Office, Service Géologique de Belgique, Rue Jenner 13,
    Brussels 1000, Belgium; RuthA@gwp.uk.com
3   Newmont Mining, 6363 South Fiddler's Green Circle Suite 800, Greenwood Village, CO 80111, USA;
    Thomas.Brady@Newmont.com
4   Institute of Environmental Sciences CML, Leiden University, Einsteinweg 2, Leiden 2333 CC,
    The Netherlands; guinee@cml.leidenuniv.nl
5   United States Geological Survey (USGS), 954 National Center, Reston, VA 20192, USA; jhammars@usgs.gov
6   Competence Center Sustainability and Infrastructure Systems, Fraunhofer Institute for Systems and
    Innovation Research ISI, Breslauer Straße 48, Karlsruhe 76139, Germany; torsten.hummen@isi.fraunhofer.de
7   Align Consulting, 1134 Cross Creek Ct., Sheridan, WY 82801, USA; andrea@alignconsultants.com
8   econsense—Forum for Sustainable Development of German Business, Oberwallstraße 24, Berlin 10117,
    Germany; l.schneider@econsense.de
9   The Life Cycle Group CyVi Institut des Sciences Moléculaires (ISM), Université de Bordeaux 1—UMR 5255
    CNRS, 351 Cours de la libération—Bât A12, TALENCE cedex 33 405, France;
    guido.sonnemann@u-bordeaux.fr
10 Lulea Technical University, Luleå 971 87, Sweden; Par.Weihed@ltu.se
*   Correspondence: drielsma@euromines.be; Tel.: +32-2-775-6305; Fax: +32-2-770-6303
†   These authors contributed equally to this work.

Academic Editors: Mario Schmidt and Benjamin C. McLellan
Received: 15 December 2015; Accepted: 16 February 2016; Published: 26 February 2016

Abstract: This paper captures some of the emerging consensus points that came out of the workshop "Mineral Resources in Life Cycle Impact Assessment: Mapping the path forward", held at the Natural History Museum London on 14 October 2015: that current practices rely in many instances on obsolete data, often confuse resource depletion with impacts on resource availability, which can therefore provide inconsistent decision support and lead to misguided claims about environmental performance. Participants agreed it would be helpful to clarify which models estimate depletion and which estimate availability, so that results can be correctly reported in the most appropriate framework. Most participants suggested that resource availability will be more meaningfully addressed within a comprehensive Life Cycle Sustainability Assessment framework rather than limited to an environmental Life Cycle Assessment or Footprint. Presentations from each of the authors are available for download [1].

Keywords: abiotic natural resources; Life Cycle Assessment; minerals; mining; ore grades; reserves; resource availability; resource scarcity; safeguard subject; raw-materials

1. Introduction

On Wednesday 14 October 2015, a global group of approximately 50 experts from academia, consulting, regulators, primary industry, down-stream sectors and standards bodies gathered at the Natural History Museum London to exchange recent findings on the way that life cycle assessment is currently applied to the use of raw-materials.

In welcoming workshop participants, the hosts explained that the road travelled in developing Natural Resources as a safeguard subject (or Area of Protection (AoP)) in Life Cycle Impact Assessment (LCIA) had been a long one, but that some more recent milestones along the way served as useful background for the day's discussion. Namely, the mining industry (Euromines and the International Council on Mining & Metals (ICMM)) had held a series of key workshops during the years 2011–2014 to bring experts from within and outside the mining industry together to discuss Life Cycle Assessment (LCA) in the context of Resource Efficiency policies. This led to greater mining industry participation in other forums such as the European Commission Joint Research Centre Workshop on "Security of supply and scarcity of raw-materials: a methodological framework for sustainability assessment" in 2012 [2] and the 55th Discussion Forum on LCA (DF-55) held in Zurich in 2014 [3].

It was announced that the industry had drafted a journal article drawing upon its experiences with LCA and the knowledge of its exploration, geology, and economic experts that is now freely available from the International Journal of Life Cycle Assessment [4].

The authors of that article suggested that development of a globally agreed upon method for assessment of abiotic raw-material inputs in Life Cycle Impact Assessment (LCIA) could be characterized by a certain amount of *confusion, resistance* and *frustration* and that, according to Knoster *et al.* [5], this quite possibly stemmed, respectively, from the lack of a common *vision* across disciplines of the potential threat or impact to measure; from a lack of aligned *incentives* amongst the different experts for developing such a method; and therefore a general lack of *knowledge* sharing and data availability between disciplines. The Workshop hosts invited the participants to begin a process of improving the sharing of knowledge and the seeking of common goals at the Workshop.

All presentations at the Workshop are available for download [1].

2. Results and Discussion

2.1. Status and Limitations: The Data

The first session of the workshop centered on the data typically drawn upon for estimating potential environmental impacts on abiotic natural resources in LCA. These data are typically generated by or for the mining industry and its (financial) stakeholders, but also for various government departments. The discussions were therefore designed to increase LCA-practitioners' familiarity with mining.

2.1.1. Economics of Resource Supply and Use

Tom Brady (Chief Economist, Newmont Mining, Greenwich Village, CO, USA) presented a visual summary of the typical process of identifying and reporting Mineral Reserves from the perspective of a mining executive. Central to his presentation was the use of the Committee for Mineral Reserves International Reporting Standards (CRIRSCO) definitions to classify different materials identified during exploration work. Mine planning (both in terms of the size and shape of the proposed mine, but also the schedule and sequence of mining) guides the identification of different classes of Mineral Resources and Mineral Reserves. Successive iterations of sampling, data interpretation and mine planning alter the estimates of each—even after a mine begins operation (Newmont typically only reports Proven Mineral Reserves once a mine has been operating as designed for 12 months). This involves consideration of several modifying factors that include processing, metallurgical, economic, marketing, legal, environmental, socio-economic and geopolitical factors. In particular, Mineral Reserve estimates fluctuate greatly as assumptions about future commodity prices change.

As of 31 December 2014, Newmont Mining Gold Reserves varied by up to 30% depending on whether a gold price of USD 1100/ounce was assumed, or USD 1500/ounce. In addition, when metal prices are high, exploration expenditure and discoveries tend to increase such that new Mineral Resources more than make up for the Mineral Reserves extracted. Whereas LCIA methods often assume that the stock of Mineral Resources and Mineral Reserves is fixed and depleting, in fact they increase or decrease with fluctuations in availability of infrastructure, exploration budgets, geological knowledge, market prices, projected production costs and technology development.

2.1.2. Resource Data: The Providers' Perspective

Jane Hammarstrom (Co-Chief, Global Mineral Resource Assessment Project, USGS, Reston, VA, USA) presented an explanation of the information services that the USGS provides, underlining how it compiles estimates of global Mineral Resources and Reserves and also explaining how it provides science-based assessments of likely Undiscovered Mineral Resources. While USGS definitions of Mineral Resources and Mineral Reserves largely match those of CRIRSCO, estimates from different sources may lack consistency owing to the different needs of, say, government and individual mining companies. Depending on the purpose and the timeframe considered, estimation methods and professional judgments may differ (e.g., commodity price and production cost forecasts). Reserve figures are estimates and they are snapshots in time that depend on several factors including demand, exploration budgets, recycling rates, technology, economics, social license to operate and environmental performance and therefore should only be interpreted together with the accompanying qualitative information provided by the Survey. Therefore, the notion that reserve figures tell us how many years remain until a natural resource is depleted must be rejected. Copper data also demonstrate the falsity of the notion that as ore is mined reserves necessarily decrease (copper reserves doubled from 1990 to 2013). It is questionable which of the different estimates of Resources and Reserves provided by the USGS could plausibly serve as a basis for measuring resource depletion. Neither the USGS Reserve Base nor the theoretical world resources is an immediately obvious or justified choice.

2.1.3. Resource Data: The Users' Perspective

Ruth Allington (Treasurer of Pan-European Reserves & Resources Reporting Committee, CRIRSCO, Brussels, Belgium) presented an overview of CRIRSCO, its aims, history, make-up and governance. The CRIRSCO-aligned definitions of Mineral Resources and Mineral Reserves have a history stretching back at least as far as the 1980s, with broad acceptance globally. The accurate and reliable reporting of mineral exploration results, Mineral Resources and Mineral Reserves is fundamental not only to mining stakeholders (for transparency of commodity markets), but also to wider society including the LCA community. CRIRSCO requirements follow some main principles related to *transparency*, *materiality*, *competence* and *impartiality*. In-particular, the role of the Competent (Qualified) Person, as required by CRIRSCO-aligned reporting codes and standards [6], is critical to upholding those principles. The Competent Person is named publicly as being personally responsible for proper estimation of Mineral Resources and Mineral Reserves and is subject to potential disciplinary action from the relevant CRIRSCO-affiliated professional organization. When the Competent Person identifies a Mineral Reserve, it must be demonstrated and this was done through a thorough analysis of the modifying factors described by others (see above) including relatively volatile socio-economic aspects such as commodity price. It is argued that CRIRSCO and its Competent Person concept are the keys to stakeholder confidence in any public reporting of LCA results based on estimates of Mineral Resources or Mineral Reserves and the LCA community should beware of embracing this economic data without acknowledging its limitations for their environmental work. To do so gives rise to misleading results and, given the CRIRSCO principles of *transparency*, *materiality*, *competence* and *impartiality*, would raise an ethical issue.

2.2. Status and Limitations: The Methods

The second session of the workshop centered on the methods typically used to estimate potential environmental impacts on abiotic natural resources in LCA. These methods have typically been developed by academics or LCA practitioners in the context of overall LCA frameworks or software. The discussions were therefore designed to increase mining professionals' familiarity with LCA.

2.2.1. Drivers for LCA of Resource Supply and Use

Andrea Russell-Vaccari (Principal Consultant, Align Consulting, Sheridan, WY, USA) introduced LCIA and safeguard subjects (Areas of Protection), their state of development and decision-makers' needs related to abiotic raw-materials. For impact category selection, the International Organization for Standardization (ISO) recommends characterization methods for the AoP be broadly agreed, environmentally relevant and describing a "distinct identifiable environmental mechanism" [7]. The resource depletion impact category is hampered by insufficient understanding of the Natural Resources AoP and hence the issue to address [8]. This is due to the variability of concepts like resource availability over space and time. Figure 1 is a visual representation of the questions that stakeholders ask about abiotic raw-materials: from environmental impacts to economic impacts; from short-term effects to long-term effects; and from micro-economics (product systems) to macro-economics (whole economies). Whereas Abiotic Depletion Potential (ADP) is easy to assess for LCIA methodology developers (top right of Figure 1), the results are not meaningful for those in the other three quadrants of Figure 1.

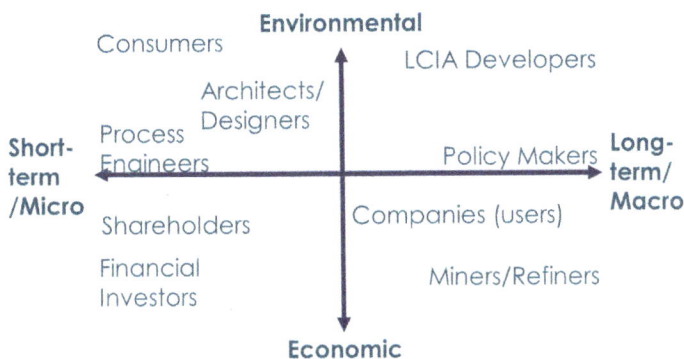

Figure 1. A proposed mapping of different decision-makers across a range of different abiotic raw material information needs (adapted from [9]).

The proposed constraint with which the assessment is made (Ultimate Reserve or Crustal Content) is so remote that it is of little relevance to decision-makers. *Availability* of materials is of more general concern. The sensitivity analysis of using other constraints, recommended by [10], demonstrates that the distinction between depletion and availability is critical, because it can actually be decisive for product selection. Because the needs of decision-makers are different to the environmental focus provided by LCIA and the AoP concept, abiotic raw-material assessment needs to expand beyond the confines of LCA and embrace other tools. LCIA alone is not able to provide adequately robust information for all decision making.

2.2.2. Abiotic Depletion Potential Method, Its Variants and Example Results

Jeroen Guinée (Assistant Profesor, CML, Leiden, The Netherlands) is one of the main authors of the original ADP method most commonly used in LCA and he presented a history of the philosophy behind it. There can be no scientifically correct method for assessing potential environmental impacts on abiotic natural resources because the concept of "resource" straddles the economic–environmental divide and none of the physical parameters or results can actually be validated empirically. This requires LCA practitioners to make a number of philosophical choices—rather than use inductive reasoning—when developing resource assessment models. Choices include a definition of the resource problem to be addressed; whether or not to consider it a problem distinct from deterioration of the environment; and which data to use. While availability, criticality, *etc.* are also of interest, the Institute of Environmental Sciences (CML) ADP method only addresses depletion of a resource in terms of reduction of the geological stocks present on Earth (Ultimate Reserve or Crustal Content). LCA results are highly sensitive to deviations from the use of Ultimate Reserves (Crustal Content). CML definitions differ from those of CRIRSCO because "resource" is commonly used to describe a category of mineral occurrences as well as to a material in its own right and is therefore ambiguous (One should perhaps rather refer to Natural Resources as natural materials or raw-materials, in order to better distinguish them from Mineral Resources as defined by CRIRSCO and USGS). The preferred parameter to describe the stock available for all of humankind would be the Ultimately Extractable Resources (or Extractable Global Resource), but this is impossible to determine because of our inability to predict what will ultimately be extracted in the future as technology advances and economic and social conditions evolve. Crustal Content is the second best parameter available and CML recommends using it as the baseline with other parameters providing an understanding of the sensitivity of the result to the choice made [8]. Unfortunately, such sensitivity analysis may not assist decision-making, because contradictory indications can often be the result. This is because USGS Reserve Base and Economic Reserve (Mineral Reserve) data involve economic considerations not directly related to resource depletion (such as structure of individual material markets, social conditions reflected in labor costs, negotiating power of mining companies and relative cost of identifying new reserves).

2.2.3. Supply as an Alternative Safeguard Subject

Torsten Hummen (Researcher, Fraunhofer ISI, Karlsruhe, Germany) introduced raw-materials criticality assessment as it is developed and practiced by the European Union. Whereas $ADP_{[Crustal\ Content]}$ looks at environmental impacts on the resource itself (depletion of natural stocks over the long term), criticality assessment looks at short-term impacts on an economy due to disturbances in a supply chain. As for classical risk management [11], criticality typically has two dimensions—a supply risk dimension and an economic importance dimension, which gives an idea of the potential consequences of a supply disruption. Criticality assessment is not yet an internationally standardized method. Criticality is tied to a particular viewpoint, *i.e.*, no raw-material is critical itself in an absolute sense, but it may be critical to somebody somewhere, under the prevailing conditions at some point in time. For example, heavy rare earth elements are considered critical for EU manufacturing in the period 2014–2024 because of their relatively high economic importance for the EU economy and the relatively high risk associated with their supply to the EU. Criticality assessment highlights current issues and informs policy or business decisions. As criticality is a relative assessment, there is no one correct place to draw the line between critical and non-critical raw-materials. In the EU, expert judgment and benchmarking with real-world markets (e.g., that of tantalum) have informed placement of the thresholds. LCA, on the other hand, is an internationally standardized methodology [7]. The environmental impact assessment of established, already employed technologies is the standard application of LCA. There are a number of questions about whether it is appropriate to integrate criticality or supply risk issues (not directly related to environmental impacts) into the Natural Resources AoP of LCA. Though Life Cycle Sustainability Assessment (LCSA) might be a more appropriate place for such assessments, current difficulties related to interdependency of criteria may increase there.

2.3. Paths Forward

Having attempted to bring delegates from different backgrounds to a shared level of basic understanding, the third session of the workshop centered on how better to employ life-cycle thinking in meeting decision-makers' needs related to abiotic raw-materials. The discussions were therefore designed to help ensure that the basis of future research is clear across the relevant disciplines.

2.3.1. Potential Paths Forward and Some Dead-Ends

Pär Weihed (Professor, LTU, Luleå, Sweden) summarized an International Journal of Life Cycle Assessment article he had recently co-authored: *"Mineral resources in life cycle impact assessment—defining the path forward"* [4]. He recommended a multi-stakeholder report titled "Breaking New Ground" [12] to all participants as an invaluable reference for assessment of potential impacts of the use of abiotic raw-materials. Current LCA models introduce unacceptable levels of uncertainty and are not working correctly as decision-making tools. "USGS data are intended to inform about each individual commodity's market conditions and can only be correctly interpreted together with the qualitative information provided in USGS commodity summaries" [4]. Harmonized terminology should be used so that LCA practitioners can build better mutual understanding with the mineral industry. Only Crustal Content data are stable and comprehensive enough to support a physical estimate of ADP—if that is desired and deemed useful. Some more promising tools are being explored within LCSA for assessing the *availability* of abiotic materials, which is defined at any given moment in time by market demand, politics, markets and technology. Observed periods of decline in average ore grades must neither be interpreted as a sign of depletion, nor as an indicator of reduced availability. In a circular economy, metals must be regarded as flows and not stocks and all five forms of capital (*i.e.*, natural, manufactured, human, social and financial capital) that are critical in the context of future abiotic raw-materials availability should eventually be considered within an LCSA tool.

2.3.2. Environmental Depletion and Economic Scarcity

Laura Schneider (Manager Environmental Issues, econsense, Berlin, Germany) presented one of the tools previously mentioned—the economic resource scarcity potential model (ESP, [13]). The ESP model allows for an assessment of raw-material availability beyond geologic finiteness and enables an identification of potential supply risks associated with the use of abiotic raw-materials. ESP defines "scarcity" as limited supply of a demanded raw-material, which could either be caused by depletion of physical stocks (long-term concern) or caused by constraints in the supply chain (short-term concern). Potential physical and economic drivers of scarcity could create direct constraints upon raw-material provision capability and environmental and social drivers could create indirect constraints. Because raw-material provision capability is potentially affected by all these constraints, it should be assessed with a multi-dimensional approach (e.g., LCSA) (here, raw-material provision capability—or supply—is the safeguard subject that can introduce risk to the studied system, whereas LCA is designed to assess the impacts of the studied system upon the external environment). Schneider *et al.* [14] further propose that ADP assessment should be extended to include anthropogenic stocks (potential secondary raw-materials) in its scope. ADP[Crustal Content] and ESP provide different information for different decisions. For example, ADP[Crustal Content] highlights the use of nickel in hybrid cars, whereas ESP highlights use of rare earths in the same hybrid cars. Comprehensive assessment of abiotic raw-material provision capability, including physical, environmental, economic and social constraints should complement and enhance current LCA practice.

2.3.3. Possibilities in Life Cycle Sustainability Assessment

Guido Sonnemann (Professor, Université de Bordeaux, ISM, TALENCE cedex, France) presented a proposed framework for integrating criticality of raw-materials into LCSA [15]. A new perspective was called for to move from assessing geological stock depletion to including security of supply aspects—as criticality assessment and some water scarcity assessment methods do. LCSA could integrate environmental, social and economic impacts of the studied system. For both water and other raw-materials, a greater geographical explicitness of sophisticated LCA models could take account of variation in climate-related and geopolitical supply risks. Sonnemann *et al.* [15] propose a framework and ISM continues to develop models for individual strands of the framework. The complementarity of criticality assessment and LCA of products suggests there is high potential for LCSA to answer stakeholders' raw-materials concerns. Data generated for LCA can provide information about abiotic raw-material use that can be usable in LCSA, but these data are more valuable the more geographically explicit they become. Supply risk is an economic problem that is perceived differently in various parts of the world and can be addressed in the full LCSA framework—not by isolated indicators hidden in an environmental LCA or footprint result (here, it is proposed that LCSA assess the studied system's contribution to increased supply risk for others, which is consistent with the general life cycle approach of assessing impacts of the studied system upon its surroundings). Using other tools to address the capability gaps left by LCA regarding raw-materials is one way to support better decision making. The Natural Resources AoP needs to be rethought by developing a multi-dimensional approach as part of LCSA—going beyond environmental LCA.

2.4. Discussion

Discussions throughout the day confirmed that there is no common understanding about what the Natural Resources AoP represents and how impacts upon it should be assessed (see also [3,8]). Part of the confusion comes from different and overlapping definitions of key terms in two key areas.

Firstly, whereas most LCIA literature refers to assessing resource *depletion* as an environmental impact, most decision-makers, stakeholders, and researchers try to understand resource *availability* as a sustainability issue that may either impinge upon the studied system or arise from it (here, we have chosen to use the definitions proposed by [4]).

Secondly, while the generators and users of Mineral Resources and Mineral Reserves data use specific definitions of each, LCA-practitioners use the term *reserve* more generally and English-speakers use the term *resource* to refer to a Mineral Resource as well as to a material in its own right.

The CML ADP[Crustal Content] method addresses resource *depletion* in terms of reductions of the Earth's geological stocks and the entire Crustal Content is not a Mineral Resource and is therefore not relevant to raw-material *availability*. On the other hand, whereas the International Life Cycle Database (ILCD) Handbook proposes that the CML ADP[Reserve Base] method be used, USGS Reserve Base data incorporate economic considerations not directly related to resource *depletion*. The group heard that the ILCD Handbook also redefined the aim of ADP to reflect *availability* of resources for human use, rather than *depletion* potential [16]. This is not trivial—depending on whether *depletion* or *availability* is assessed in an LCIA, one is either dealing with an environmental or an economic issue and workshop participants had seen during the day that contradictory material or product selections can and do result.

Apart from this, there is the problem that it has not been possible to implement the ILCD Handbook recommendation because the USGS had ceased estimating Reserve Base when the Handbook was adopted. This means that calculations of ADP[Reserve Base] now rely on obsolete data. After the US Bureau of Mines closed, the USGS discontinued its Reserve Base estimates. Each annual USGS Mineral Commodity Summary includes a discussion of world resources for a given commodity. From 1996 until 2009, Reserve Base estimates were only updated with changes in reserves because available updates on the non-reserve component of the Reserve Base were insufficient to support defensible Reserve Base estimates. Therefore, starting with Mineral Commodity Summaries 2010, publication of a Reserve Base was discontinued [17].

Application of ADP[Reserve Base] requires regular updating of characterization factors because Mineral Resources and production change over time. The group discussed how the lack of such updates had serious ramifications for the current Product Environment Footprint (PEF) pilot project of the EU, as it had adopted the ILCD Handbook recommendation in its rules for participants. One of the European Union's intended uses of its PEF Guide and relevant PEF category rules is to support claims about the *environmental* superiority or equivalence of a product when compared to others that perform the same function [18]. If results using $ADP_{[Reserve\ Base]}$ only are used in this way based on obsolete data, it cannot be considered good practice and will certainly lead to misguided claims about environmental performance. To illustrate the dangers of miscommunicating about environmental performance, one participant raised the possibility that a bridge designed to minimize its environmental footprint, could soon afterwards be assessed with a sub-optimal footprint due only to prevailing economics changing $ADP_{[Reserve\ Base]}$ characterization factors—not because of any change in the environmental performance of the bridge. Should the bridge then be replaced?

Given the consensus on the above aspects, different views were expressed as to the usefulness of the $ADP_{[Crustal\ Content]}$ for assessing *availability* of raw-materials. Some academics argued that partial *depletion* of geological stocks could contribute to reduced *availability* of raw-materials over the very long term. That is, a raw-material may become less available to future generations even if functionally it is replaced by other materials or technologies. Geologists remained concerned that if *availability* is to be assessed, Crustal Content is an inappropriate yardstick to use because much of it can safely be assumed to be unavailable. Most participants seemed to agree that *availability* was an economic issue more meaningfully addressed within a comprehensive LCSA framework.

At the request of CML, it was agreed that in future discussions it should always be specified which particular CML option for ADP calculation is being referred to. The ILCD Handbook of the European Commission Joint Research Centre contradicts CML Guidance by specifying use of USGS Reserve Base data instead of Crustal Content data. When this particular ILCD Handbook recommendation is criticized, it should be clearly stated rather than criticizing the whole ADP method as published by CML.

Some extra suggestions were made to advance the search for a suitable tool to assess *availability* of raw-materials. The available quantity of raw-materials at any given time is always a subset of the total natural occurrence of that material. Therefore, there was some agreement across disciplines that available quantities of raw-materials (e.g., any class of Mineral Resources or Mineral Reserves) should be modeled as fund resources (rather than stock resources) and that dissipative outflows from studied systems—rather than inflows to them—were the concern to address in order to maximize continued *availability* of raw-materials. This would also give recognition to the fact that raw-materials are not always consumed or dissipated, but often remain available as an anthropogenic source for recycling. It may prove more practical and relevant to use data on dissipative outflows as a proxy for reduced *availability* (See also [3]).

Finally, although several participants were working on stand-alone frameworks to include assessment of raw-material *availability* and supply risks, it was acknowledged that a tool-box approach may prove to be more appropriate and that eventually only one set of globally agreed tools should be promoted as good practice in the field.

3. Conclusions

Over the last 20 years, the idea that Natural Resources exist separately from Human Health and the Natural Environment has been maintained in LCA theory and yet, as this workshop showed, the idea is not a simple one. Dictionary definitions of the term *resource* invariably refer to *availability* to or *use* by humans. Some contend that Natural Resources are components of the Natural Environment seen through a utilitarian lens. It is easier to understand that *health* and *environment* obey their own fundamental laws (*i.e.*, the laws of thermodynamics and quantum mechanics) and exist in some form without human interaction. Such a distinct set of environmental mechanisms is missing for Natural Resources, which makes them difficult to characterize or evaluate within the logical construct of LCIA.

To better understand each other, delegates at this Workshop often found it necessary to talk instead of *raw-materials*. As defined by leading geological institutions, Mineral Resources and Mineral

Reserves are snapshots of raw-material *availability* at a given point in time given prevailing market conditions. They do not represent the total stock of material available to humankind for all time and they are not necessarily diminished by the mining of ore. It is impossible to predict future human capabilities and therefore how much of a raw-material will ultimately prove extractable from the Earth.

CML therefore considers that Crustal Content is the best available baseline to determine resource *depletion* and recommends using it as such in LCIA. European Commission Joint Research Centre Guidance addresses materials in terms of *availability* for human use and suggests ADP[Reserve Base] should be used to express that. However, USGS Reserve Base data are obsolete (no alternative source of Reserve Base data exists) and largely underestimate long-term raw-material *availability* for human use. If used to support claims about the environmental performance of a product or organization, ADP[Reserve Base] will certainly lead to misguided claims being made.

The weight of available evidence suggests that extraction of natural resources from the environment is not decreasing foreseeable *availability* of natural materials for human use. Even observed periods of decline in average ore grades have been accompanied by significant increases in Mineral Reserves (See also [4,19,20]). Such ore grade trends can neither be interpreted as a sign of depletion, nor as indicating reduced *availability*. Raw-material *availability* and the associated supply risks are overwhelmingly the result of economic, technological and geo-political forces and most participants seemed to agree that such an AoP will be more meaningfully addressed within a comprehensive LCSA framework rather than limited to an environmental LCA or Footprint.

To better define how life cycle thinking can help answer questions related to the extraction and use of abiotic raw-materials, it would be helpful to clearly distinguish *availability* from *depletion* and clarify which of these is estimated by each of the existing LCIA methods, so that results can be correctly reported in the most appropriate framework. It could also be a helpful convention to refer to Natural Resources used in studied systems as *raw-materials* and to clearly distinguish them from Mineral Resources and Mineral Reserves as defined by CRIRSCO and USGS.

Without a shared vision, agreed terminology and common incentives, collaboration across disciplines on assessment of impacts of abiotic raw-material inputs to production and consumption systems will continue to be very difficult. The only satisfactory path forward is more dialogue between experts in the fields of LCA, commodities trading, environmental science, exploration geology, industrial ecology, technological innovation, metallurgy, mineral economics, minerals policy, product policy and sustainable development.

Acknowledgments: The authors would like to thank all the workshop participants for their contributions. Additionally, we would like to acknowledge the European Copper Institute, the European Association of Mining Industries Metal Ores and Industrial Minerals (Euromines), the International Council on Mining & Metals (ICMM), the Natural History Museum London and the Nickel Institute for their financial and technical support during the organization of the workshop.

Conflicts of Interest: The authors declare no conflict of interest.

References

1. Mineral Resources in LCIA: Mapping the path forward. Available online: http://www.euromines.org/events/2015-10-14-mineral-resources-lcia-mapping-path-forward (accessed on 19 October 2015).
2. Mancini, L.; de Camillis, C.; Pennington, D. *Security of Supply and Scarcity of Raw Materials. Towards a Methodological Framework for Sustainability Assessment*; Publications Office of the European Union: Luxembourg, Luxembourg, 2013.
3. Vadenbo, C.; Rorbech, J.; Haupt, M.; Frischknecht, R. Abiotic resources: New Impact Assessment Approaches in View of Resource Efficiency and Resource Criticality—55th Discussion Forum on Life Cycle Assessment, Zurich, Switzerland, 11 April 2014. *Int. J. Life Cycle Assess.* **2014**, *19*, 1686–1692. [CrossRef]
4. Drielsma, J.A.; Russell-Vaccari, A.J.; Drnek, T.; Brady, T.; Weihed, P.; Mistry, M.; Perez Simbor, L. Mineral resources in life cycle impact assessment—Defining the path forward. *Int. J. Life Cycle Assess.* **2016**, *21*, 85–105. [CrossRef]

5. Knoster, T.; Villa, R.; Thousand, J. A framework for thinking about systems change. In *Restructuring for Caring and Effective Education: Piecing the Puzzle Together*; Villa, R., Thousand, J., Eds.; Paul H Brookes Publishing Co.: Baltimore, MD, USA, 2000; pp. 93–128.
6. Committee for Mineral Reserves International Reporting Standards: National Reporting Standards. Available online: http://www.crirsco.com/national.asp (accessed on 19 October 2015).
7. International Standardization Organization (ISO). *Standard 14044: Environmental Management—Life Cycle Assessment—Requirements and Guidelines*; ISO: Geneva, Switzerland, 2006.
8. Hauschild, M.Z.; Goedkoop, M.; Guinée, J.; Heijungs, R.; Huijbregts, M.; Jolliet, O.; Margni, M.; de Scrhyver, A.; Humbert, S.; Laurent, A.; *et al.* Identifying best existing practice for characterization modelling in life cycle impact assessment. *Int. J. Life Cycle Assess.* **2013**, *18*, 683–697. [CrossRef]
9. Russell-Vaccari, A.J. Global Drivers for LCA of Resource Supply & Use. Available online: http://www.euromines.org/system/files/events/2015-10-14-mineral-resources-lcia-mapping-path-forward/6-drivers-and-needs-vaccari-2015-workshop.pdf (accessed on 19 October 2015).
10. Van Oers, L.; de Koning, A.; Guinée, J.B.; Huppes, G. *Abiotic Resource Depletion in LCA: Improving Characterisation Factors for Abiotic Resource Depletion as Recommended in the New Dutch LCA Handbook*; Road and Hydraulic Engineering Institute, Ministry of Transport and Water: Amsterdam, The Netherlands, 2002.
11. International Standardization Organization (ISO). *Standard 31000: Risk Management—Principles and Guidelines*; ISO: Geneva, Switzerland, 2009.
12. International Institute for Environment and Development (IIED). *Breaking New Ground: The Report of the Mining, Minerals, and Sustainable Development Project*; Earthscan Publications Ltd.: London, UK, 2002.
13. Schneider, L.; Berger, M.; Schüler-Hainsch, E.; Knöfel, S.; Ruhland, K.; Mosig, J.; Bach, V.; Finkbeiner, M. The economic resource scarcity potential (ESP) for evaluating resource use based on life cycle assessment. *Int. J. Life Cycle Assess.* **2013**, *19*, 601–610. [CrossRef]
14. Schneider, L.; Berger, M.; Finkbeiner, M. Abiotic resource depletion in LCA—Background and update of the anthropogenic stock extended abiotic depletion potential (AADP) model. *Int. J. Life Cycle Assess.* **2015**, *20*, 709–721. [CrossRef]
15. Sonnemann, G.; Gemechu, E.D.; Adibi, N.; De Bruille, V.; Bulle, C. From a critical review to a conceptual framework for integrating the criticality of resources into life cycle sustainability assessment. *J. Clean Prod.* **2015**, *94*, 20–34. [CrossRef]
16. European Commission Joint Research Centre. *ILCD Handbook. Recommendations Based on Existing Environmental Impact Assessment Models and Factors for Life Cycle Assessment in European Context*; IES, Joint Research Centre: Ispra, Italy, 2011.
17. US Geological Survey. *Mineral. Commodity Summaries 2010*; U.S. Department of the Interior: Reston, VA, USA, 2010.
18. European Union. Commission recommendation of 9 April 2013 on the use of common methods to measure and communicate the life cycle environmental performance of products and organisations. *Off. J. Eur. Union.* **2013**, *56*, 1–210.
19. Schodde, R. The Key Drivers behind Resource Growth: An Analysis of the Copper Industry over the Last 100 Years. Available online: http://www.minexconsulting.com/publications/ Growth%20Factors%20for% 20Copper%20SME-MEMS%20March%202010.pdf (accessed on 29 January 2016).
20. West, J. Decreasing metal ore grades: Are They Really Being Driven by the Depletion of High-Grade Deposits? *J. Ind. Ecol.* **2011**, *15*, 165–168. [CrossRef]

resources

MDPI

Article

Mineral Resources: Reserves, Peak Production and the Future

Lawrence D. Meinert *, Gilpin R. Robinson Jr. and Nedal T. Nassar

U.S. Geological Survey, 12201 Sunrise Valley Drive, MS 913, Reston, VA 20192, USA;
grobinso@usgs.gov (G.R.R.); nnassar@usgs.gov (N.T.N.)
* Correspondence: Lmeinert@usgs.gov; Tel.: +1-703-648-6100

Academic Editor: Mario Schmidt
Received: 22 December 2015 ; Accepted: 16 February 2016 ; Published: 29 February 2016

Abstract: The adequacy of mineral resources in light of population growth and rising standards of living has been a concern since the time of Malthus (1798), but many studies erroneously forecast impending peak production or exhaustion because they confuse reserves with "all there is". Reserves are formally defined as a subset of resources, and even current and potential resources are only a small subset of "all there is". Peak production or exhaustion cannot be modeled accurately from reserves. Using copper as an example, identified resources are twice as large as the amount projected to be needed through 2050. Estimates of yet-to-be discovered copper resources are up to 40-times more than currently-identified resources, amounts that could last for many centuries. Thus, forecasts of imminent peak production due to resource exhaustion in the next 20–30 years are not valid. Short-term supply problems may arise, however, and supply-chain disruptions are possible at any time due to natural disasters (earthquakes, tsunamis, hurricanes) or political complications. Needed to resolve these problems are education and exploration technology development, access to prospective terrain, better recycling and better accounting of externalities associated with production (pollution, loss of ecosystem services and water and energy use).

Keywords: mineral resources; peak copper; sustainability; limits to growth; reserves; production; depletion; life cycle assessment; material flow; ecosystem services

1. Introduction

At least since the time of Malthus (1798) [1], there has been concern about the adequacy of resources to support a growing human population on planet Earth. In more recent times, there has been an ebb and flow between studies that predict the near-term exhaustion or peak production of resources [2–13] and those that are more optimistic [14–20]. The one thing that is common among every prediction about the exhaustion or production peak of resources by a particular date is that they have been wrong. This is well illustrated, but by no means proven or vindicated, by the famous bet between Julian Simon and Paul Erlich about the price of five benchmark metals, chromium, copper, nickel, tin and tungsten, ten years in the future [21]. Even though Dr. Erlich was quite confident that prices would have to be higher in the future due to resource scarcity, he lost that bet because prices were actually lower, both in real and inflation-adjusted dollars. However, had the bet been made at a different time, the outcome could have been the opposite of what transpired. Therefore, what is it that leads to such uncertainty about what seems to be a straightforward argument?

On the one hand, there is no disputing that Earth has a finite mass and if resources are consumed at some definable rate, then eventually they will be used up [22]. On the other hand, we have not yet run out of any mineral commodity, and exploration and technology have more than kept up with changing demands for mineral resources throughout human history. As was said by Sheikh

Zaki Yamani, "The Stone Age did not end for lack of stone..." (quoted in [23]) How do we reconcile such contradictions?

Perhaps a useful analogy for understanding this seeming paradox between the unarguably finite amount of mineral resources on Earth and the continuing ability of society to meet its resource needs is the continuous increase in human performance in athletic endeavors, even though elementary logic dictates that it cannot continue forever. It is a demonstrable fact that new world records continue to be set in every athletic event despite the real limits to human achievement: humans will never be able to run faster than a rifle bullet, except in comic books and movies.

Although many sports analogies could be considered in this regard, running might be the most useful because the records are easily quantifiable and because most people have run at some point in their lives, even if not at a competitive level. A running event that illustrates this logic conundrum is the mile, because the 4-min benchmark was long considered unreachable. There are many excellent literary and historical accounts of the quest to break this once unbreakable mark [24]. That pace today would not even qualify for entry into the Olympics, and the current world record for the mile is 3:43.13, set in 1999 by Morocco's Hicham El Guerrouj. However, any particular record is not the point. If a 4-min mile is not the limit, is 3:40, 3:30 or 3:00? History says that no record lasts forever, yet elementary physiology tells us that improvement cannot continue indefinitely. Therefore, although obviously this is not a perfect analogy for the ultimate limits to mineral resources, it does illustrate the difficulty in reconciling finite limits with the absence of evidence of reaching those limits. That is the central theme of the present paper.

2. Background

Before discussing a framework for thinking about limits to mineral resource availability and consequences of use, it is useful to first explore the important terms, reserves and resources [11]. These terms are poorly understood by many and widely misused, particularly in the well-known *Limits to Growth* study of Meadows *et al.* [2] and more recent incarnations [10,25,26]. Reserves have a defined meaning as codified by several widely-used standards, such as the Australian Joint Ore Reserves Committee (JORC) [27], the South African Mineral Resource Committee (SAMREC) [28] and Canada's National Instrument 43-101 [29]. Furthermore, the Committee for Mineral Reserves International Reporting Standards (CRIRSCO) [30] has harmonized the various country-level codes into a common international reporting standard. A summary of various resource and reserve classifications is outlined in Table 1.

Table 1. Mineral resource and reserve classifications.

Guidelines	Resource/Reserve Classification
U.S. Geological Survey	U.S. Bureau of Mines and U.S. Geological Survey, 1980, Principles of a resource/reserve classification for minerals: U.S. Geological Survey Circular 831, p. 5 [31]
International	CRIRSCO (the Committee for Mineral Reserves International Reporting Standards), a committee of the National Mineral Reserve Reporting Organizations of Australia, Canada, Chile, South Africa, USA, U.K./Ireland and Western Europe has prepared the CRIRSCO International Reporting Template drawing on CRIRSCO-style reporting standards, the JORC Code (Australasia), SAMREC Code (South Africa), Reporting Code (U.K./Ireland/Western Europe), Canadian Institute of Mining, Metallurgy, and Petroleum (CIM) Definition Standards and Guidelines (Canada), Society for Mining, Metallurgy, & Exploration (SME) Guide (USA) and Certification Code (Chile). (URL: http://www.crirsco.com) [30]
United Nations	United Nations Economic Commission for Europe, 2003, Framework Classification for Fossil Energy and Mineral Resources: United Nations Economic Commission for Europe, Geneva [32]

Table 1. *Cont.*

Guidelines	Resource/Reserve Classification
Australia	Joint Ore Reserves Committee of The Australasian Institute of Mining and Metallurgy, Australian Institute of Geoscientists and Minerals Council of Australia (JORC), 2004; Australasian Code for Reporting of Exploration Results, Mineral Resources and Ore Reserves: The Australasian Institute of Mining and Metallurgy [27]
Canada	CIM Standing Committee on Reserve Definitions, 2004, CIM Definition Standards on Mineral Resources and Mineral Reserves: The Canadian Institute of Mining, Metallurgy and Petroleum [29]
Chile	Institute of Mining Engineers of Chile (IIMCh), 2004; Code for the Certification of Exploration Prospects, Mineral Resources and Ore Reserves, 2004: Institute of Mining Engineers of Chile [33]
China	The State Bureau of Quality and Technical Supervision, 1999 Classification for Resources/Reserves of Solid Fuels and Mineral Commodities (National Standards GB/T17766-1999): Beijing: Chinese Standard Publishing House [34]
Peru	Joint Committee of the Venture Capital Segment of the Lima Stock Exchange, 2003, Code for Reporting on Mineral Resources and Ore Reserves: Bolsa de Valores de Lima (Lima Stock Exchange) [35]
South Africa	South African Mineral Resource Committee (SAMREC), 2000, South African Code for Reporting of Mineral Resources and Mineral Reserves (The SAMREC Code): South African Institute of Mining and Metallurgy [28]
Soviet Union, CIS, COMECON countries	Diatchkov, Sergei, 1994, Principles of classification of reserves and resources in the CIS countries: Mining Engineering, v. 46, No. 3, p. 214–217 [36] Jakubiak, Z., and Smakowski, T., 1994, Classification of mineral reserves in former Comecon countries: Geological Society of London, Special Publications, v. 79, p. 17–28 [37]
United States	Legally-required system for reporting reserves by publicly-traded companies Securities and Exchange Commission—Description of Property by Issuers Engaged or to Be Engaged in Significant Mining Operations Guide 7 [38] **Proposed classification** U.S. Securities and Exchange Commission (SEC) Reserves Working Group of SME Resources and Reserves Committee, 2005, Recommendations Concerning Estimation and Reporting of Mineral Resources and Mineral Reserves, Prepared for Submission to the United States Securities and Exchange Commission: The Society for Mining, Metallurgy and Exploration, Inc. [39]

In simple terms, a reserve is a known quantity of a resource as established by drilling and sampling; it typically is expressed as X tons of material with an average grade of Y at a cutoff grade of Z. This results in a calculated amount of contained metal that potentially could be recovered, based on assumptions of cost, price and technology. Resource is a broader and more general term than reserve and includes identified material that may be less well characterized, possibly of lower grade and less certain to be economically recoverable. Resources can be converted to reserves by additional drilling or changes in economic factors, such as price or technology [40]. However, it is very important to understand that neither reserves nor resources are the same as "all there is".

This is the fundamental flaw of many studies, such as Meadows *et al.* [2], that assumed reserves, or some multiplier thereof, were "all there is" and that by applying a given annual rate of consumption, one could model how long the resource would last before disaster ensued (a more detailed analysis is given by [19,41]). A little appreciated fact is that increasing world population and standards of living lead to more production, which, in turn, requires larger reserves to sustain that production; thus, world reserves of almost all commodities are larger now than they were 50 or 100 years ago [42]. This is because the time value of money makes it uneconomic to spend unlimited amounts to convert all identified or undiscovered resources into reserves. In practice, most companies do not drill out more than about 20 or 30 years' worth of reserves, and some large mines have had ~20 years' worth of reserves for more than a century as continuing exploration proves additional reserves. Thus, modeling of how long reserves might last is fraught with the same difficulty as modeling how long the food in one's refrigerator might last: even though you keep eating three meals a day, there is still food in the refrigerator.

It is worth repeating that reserves are an economically-defined quantity and never have and never will equate to "all there is". In spite of this reality, a recent article went so far as to assert that the majority of Earth's reserves have already been consumed: "80% of the world's mercury reserves, 75% of its silver, tin, and lead, 70% of gold and zinc, and 50% of copper and manganese had already been processed through human products" [26] (p. 239), even though it is a simple fact that the world's reserves of these elements continue to evolve with improved geological assurance, technological advances and economic conditions and are presently as large as they have ever been [43].

Another way of approaching this question of how long resources will last is the "peak" concept as applied to petroleum resources by M. K. Hubbert [13,44]. It is based on an empirical observation about U.S. oil fields that production follows a bell-shaped curve of a normal distribution, and thus, the peak of production occurs when roughly half of the resource has been extracted. Thus, changes in production are used to model the entire extractable resource. Such an approach is based on many assumptions, the most important of which are that prices and technology are relatively stable. Hubbert's prediction that peak oil production for the conterminous United States would occur in the early 1970s was a reasonable extrapolation of trends leading up to the supposed "peak oil" and has been cited by many as proof that such production trend modeling is accurate [5,6,45]. However, the recent application of new and improved drilling technology (horizontal drilling and fracking) has shown that the assumptions of the Hubbert peak oil methodology were restricted to "conventional" petroleum production and have not accurately tracked the large increases in both production and reserves of petroleum and natural gas that have recently resulted from these new technologies (Figure 1) [46].

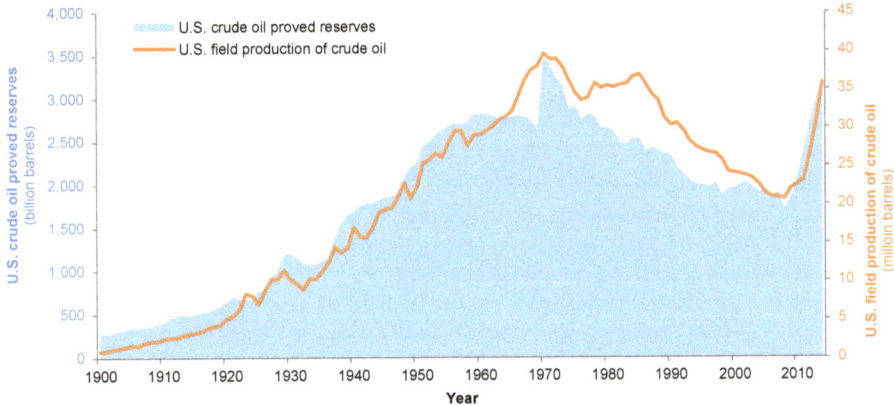

Figure 1. U.S. crude oil reserves and production showing apparent "peak oil" for conventional petroleum reserves in 1972 and growth in reserves in the past ~10 years as a result of the application of new drilling technology. U.S. barrels are equivalent to approximately 0.159 cubic meters. Data source: [47].

The point of the present paper is not to assess previous studies of resource adequacy, but rather to provide a conceptual framework for thinking about how resources, particularly mineral resources, are explored for, discovered and put into production in order to satisfy the continuing needs of society. An underlying concept is that mineral resources have been the foundation of civilization throughout history [12,48–50] and that the adequacy of such resources for future generations remains one of the central issues of our time [51,52]. Although many mineral resources could be used as examples, the below discussion will focus on copper, because it has been used throughout history; it is a major ingredient in modern construction and technology; its geologic occurrence is relatively well known; and it has a rich historical record of production data. In addition, there have been numerous studies [53,54] about copper, as well as recent predictions of "peak copper" occurring within the next 20–30 years [9,10,55].

3. Copper

Historic mine production has recovered more than 658 million metric tons (Mt) of copper (Cu) from 1700 to 2015 [43,56,57]. Of this total, over 45 percent has been produced during the past 20 years, and approximately 26 percent has been produced during the previous 10 years. In other words, roughly one quarter of all of the copper mined throughout human history has been produced in just the past 10 years (Figure 2). These percentages suggest that global copper production has doubled every 25 years in order to accommodate trends in global copper consumption.

A research group at Monash University in Australia analyzed current estimates of copper resources based on published company reports and tabulated a total of 1781 Mt of Cu contained within a total of 730 projects, with a further 80.4 Mt of Cu in China from an unknown number of deposits, yielding a world total of 1861 Mt of Cu [53]. This is very similar to a recent U.S. Geological Survey (USGS) estimate of identified Cu resources of 2100 Mt of Cu [54]. To reiterate a previous point, this 2100 Mt of Cu is the amount of identified resources, and further drilling or possible changes in prices/technology will almost certainly add to the total. However, neither reserves nor resources are "all there is". Before addressing the questions of what is known about, and the potential adequacy of, "all there is", it is useful to compare current estimates of reserves to current and projected rates of production.

Modeling based on identified resources has led some researchers to suggest that mine production of copper will peak and begin to decline within decades due to increasing demand and resource depletion [8,55]. Is this realistic? The U.S. Census Bureau estimates that world population will grow from 7.3 billion people in 2015 to 9.4 billion in 2050 [58]. At current rates of global per capita production of copper, estimated at about 2.6 kg/year in 2015, the production of primary (mine-produced) copper would grow to 24.2 Mt of copper in 2050 from the level of 18.7 Mt in 2015 [43].

However, assuming a global per capita production value of 2.6 kg/year does not take into account regional variations in per capita consumption and how they are changing due to changes in affluence or lifestyles. Nor does it take into account the increasing quantities of copper that will likely become available for recycling and thus offset the need for primary copper. Studies of various countries (e.g., [59–62]) show that per capita consumption of copper is related to the level of economic development. Per capita consumption of copper is relatively stable over time in high-GDP countries, such as the United States and Japan, at about 6 kg/year [63]. Very low levels of consumption occur in countries with low economic activity and low income levels, whereas increasing levels of per capita consumption are observed in rapidly developing countries, such as Indonesia and China [63].

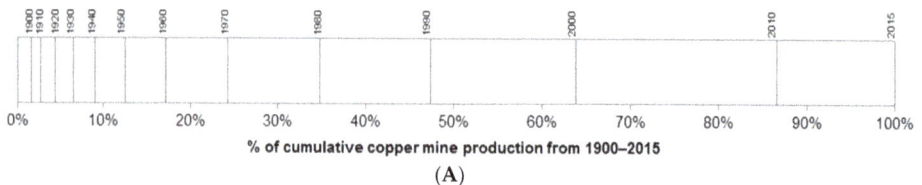

(A)

Figure 2. *Cont.*

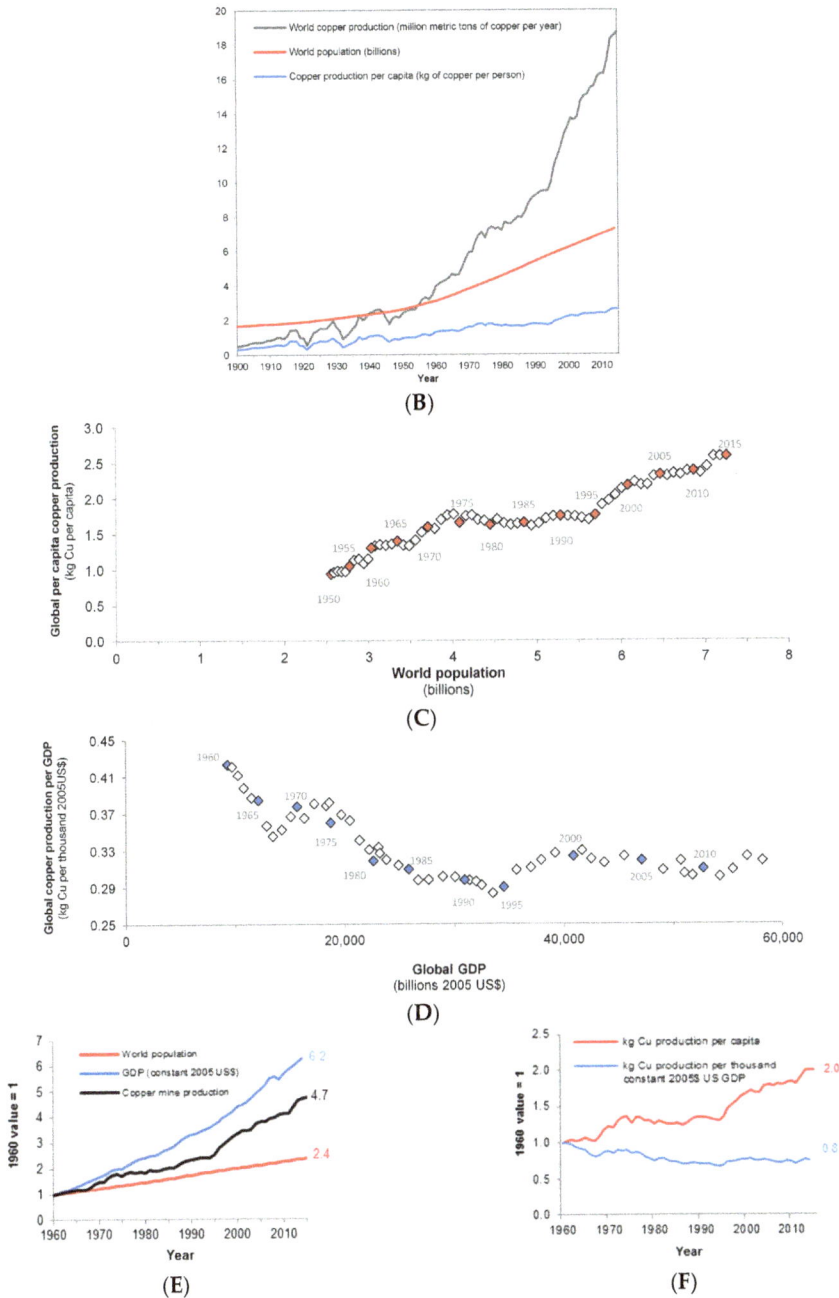

Figure 2. Trends in cumulative (**A**); annual (**B**); per capita (**C**); per constant 2005 US$ gross domestic production (GDP) (**D**); copper production on an absolute basis and normalized to one for values in year 1960 (**E,F**). Data sources: [56,57] for copper production; [58] for world population; [64] for GDP data. Data presented up to year 2015, except for those involving GDP for which the year 2014 is the most recently available.

Factoring in projected growth in population and global per capita GDP to 2050 provides a forecast of copper consumption that needs to be satisfied by mine production of 30.4 Mt of Cu. These projections indicate that between 750 and 990 Mt of primary copper will be required to satisfy projected global demand from 2012 to 2050. The projected range of estimated demand considers modest changes in the amount of copper recycling, which has increased in recent years [65]; but does not account for new technology-driven changes in demand patterns, substitution by alternative materials or other factors that could alter future demand growth for copper. To restate this, we can estimate amounts and project trends, but we cannot know the actual mix of mineral resources needed 30–50 years in the future any more than someone in 1970 could have predicted today's high-tech need for a variety of specialty metals, such as rare-earth elements.

Using these assumptions and constraints, the current estimate of 2100 Mt of copper in identified resources in major copper deposits worldwide is about twice as much copper as is estimated to be needed through 2050. Even without considering the almost certain expansion of reserves that will take place with future exploration and drilling, this makes it very unlikely that copper production will peak by 2040 due to resource depletion (Figure 3), as predicted by the studies of [9,10,55].

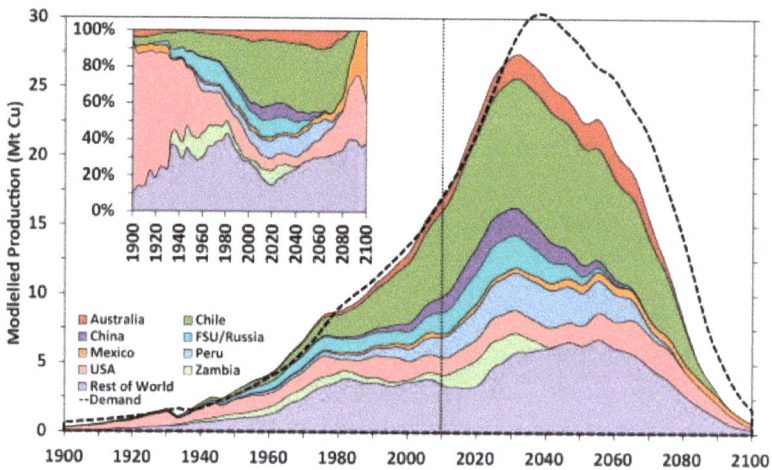

Figure 3. World copper production: historical and projected based on the modeling in [9].

Furthermore, because many of these studies dating back to Meadows *et al.* [2] confuse reserves with "all there is", their conclusions about reserve exhaustion, whether substantiated or not, do not really address the underlying question of the adequacy of mineral resources for future generations. To best address this important question, we need to know future world demand (net of recycling) for minerals, estimate as well as we can the amount of identified resources and assess the location, quality, quantity and the environmental and economic aspects of the extraction of undiscovered resources. This is a daunting task that has never been seriously undertaken for any major commodity. The best effort for copper is a recently completed global assessment of the major sources of onshore world copper resources in two, but not all, of the most important types of copper-rich deposits by the U.S. Geological Survey (USGS) [54].

Before addressing the implications of that study for the questions of "peak copper" and the adequacy of copper resources for future generations, it is important to realize what is not included in that assessment. The USGS study focused mainly on deposits to a depth of one kilometer and did not include resources in the ocean basins that cover more than 75% of the Earth's surface, nor the possibility for utilization of resources from space. Some studies have indicated that seafloor resources [66] may be as much or greater than onshore resources and others have proposed direct recovery of metals from seawater, which if economically viable, would likely eclipse all onshore resources [67]. Neither of these ideas about potential resources has been developed sufficiently to know if they are even possible, but their relevance to the present discussion is simply that it is not possible to predict the exhaustion of mineral resources when only considering part of the potential resource.

Also important is the fact that the USGS study was constrained to current prices and technology for considering what might be an economic mineral resource. If prices were 10-times higher or technology 10-times more efficient (both of which have taken place in the history of resource extraction [42]), then the assessed copper resources would be considerably higher than the presented estimates. Thus, the undiscovered copper resources estimated by Johnson *et al.* [54] are still a small subset of "all there is", but at least this estimate is closer to an ultimate resource than the identified resources categorized by Mudd *et al.* [53] and similar studies.

The USGS global copper resource assessment [54] estimated that more than 3500 Mt of copper are present in undiscovered porphyry and sediment-hosted continental copper deposits; the majority of these resources are located in Asia, followed by the Americas. Of this 3500-Mt estimate of in-ground copper resources, approximately 2000 Mt of copper is estimated as economic to recover under current economic conditions and mining technology.

A significant fraction of these undiscovered resources is estimated to occur deep beneath covering rock and sediment and commonly in remote areas. These deposits will be difficult and time consuming to discover, and many potentially economic deposits in remote and deep settings may not be discovered and developed before 2050. An estimate of undiscovered resources that are most likely to be discovered and characterized by 2050 is approximately 1100 Mt of copper [68].

Using the production rates (16.9–30.4 Mt/y of copper) previously discussed for current reserves, the range of undiscovered copper resources by the USGS study of 1100–3500 Mt of copper (keeping in mind the previously discussed caveats about this being a minimum estimate) could last more than a century. This would be in addition to the currently identified resources, which by themselves are estimated to be twice the amount needed through 2050. Therefore, the combined total is likely to last at least through the end of this century and possibly the century after that.

Kesler and Wilkinson [69] used a different, largely hypothetical, approach to estimating world copper resources. Their tectonic diffusion model predicts world copper resources of 89 billion metric tons (Bt) of copper within mineable depths, an amount that would last for millennia. Thus, although it is true that copper is not an infinite resource on planet Earth, based on current knowledge, a sufficient amount exists to fulfill the needs of at least several generations.

Of course, copper, like any other mineral resource, is not really "consumed" in the same way that energy loses its ability to perform useful work when it is "consumed." Thus, discussions of mineral resource exhaustion are implicitly referring to the lack of availability of resources in the ground and not their lack of availability in an absolute sense given that many metals are available for multiple cycles of use and re-use. Within this paradigm of ultimate resources, it is useful to discuss how much of a resource is currently in use, how much might be needed in the future to fulfill the services desired by an increasingly affluent global population and how much of what is in the ground can be put into use to satisfy these requirements [7,70,71].

Therefore, instead of worrying about reaching "peak" production or "exhausting" a resource, we should instead be more concerned about what we do with the resource after it has been extracted. Is it being used in dissipative applications [72]? How much is being "lost" (*i.e.*, placed in a state where it is prohibitively difficult or costly to recover) at each life cycle stage? Further concerns are the externalities (pollution, loss of ecosystem services and water and energy use) associated with resource production. A full life cycle assessment and material flow accounting of mineral resources would be a more fruitful conversation than just peak production due to resource exhaustion.

4. The Future

All of the preceding discussion boils down to when, but not if, a given quantity of resource will be exhausted. However, this begs the question of, "what then?" This is not as dramatic as it may seem because the common analogy between resource exhaustion and falling off a cliff is not appropriate. Resource use does not proceed full bore until the last unit of an element is consumed and then disaster unfolds. Rather, resource use follows the basic laws of supply and demand. As resources are consumed, scarcity will drive prices up, which in turn will affect demand, consumption and production. To greatly simplify, if the equilibrium price of a commodity rises ten-fold, then this will both increase supply, because previously marginal or unknown deposits will be put into production, and decrease demand, because some uses of that commodity will no longer be economic, and either production will decrease, thrifting (reducing unit consumption) will occur or other materials will be substituted. This will gradually reduce the need for the commodity in question in the long term and bring production in line with the available reserves, even though short-term supply disruption or imbalance is possible. To return to an earlier analogy, "the Stone Age did not end because we ran out of stone", we do not know which metals will be in demand 30–50 years from now; changes in technology and lifestyle may mean that society's future needs are very different from today's.

This does not mean that we should be unconcerned about resource adequacy. We should. However, we should not confuse the adequacy of mineral resource supply with the actions needed to address the needs of future generations for mineral resources.

The main resource issue for the future likely will be the development of capacity to discover and produce additional resources. There may well be enough undiscovered copper to meet global needs for the foreseeable future, but rates of exploration, discovery success and mine development will need to increase to supply the needed copper resources. Cost-effective technology will need to be developed to discover additional mineral deposits in new, deeper and concealed settings and to extract and recover resources from earth materials, while minimizing environmental impact. This technology will only be effective if it can be applied to lands where new resources may be discovered and mines developed. Nature, not society, controls where mineral concentrations occur, but society determines whether to mine or not, and land access becomes more difficult as population increases.

As much as recycling [65] and substitution [73] will be part of the solution, they cannot by themselves solve the problem. Population growth and rising standards of living, combined with the sequestration of elements, like copper in buildings, cars, cell phones, *etc.*, for periods of years to decades and, in some cases, centuries will require new primary supplies of mineral resources. The lead time from discovery to mine development can be 10–30 years. Extensive mineral exploration will be required to meet this future resource demand, because many of the undiscovered deposits will be harder to find and more costly to mine than near-surface deposits located in more accessible areas.

At a global level, it is not clear that society is making the investments in education, research and development to ensure that these mineral resources will be available for future generations [74]. The number of universities that teach and do research on mineral resources, mining engineering and metallurgy is decreasing rather than increasing, and this is paralleled by decreases in governmental funding to support geoscience research (Figure 4). National, state and provincial geological surveys and the knowledge infrastructure that they create, manage and publically provide are instrumental in the discovery process, particularly by conducting modern geologic mapping and related geochemical and geophysical surveys to identify favorable geologic environments likely to contain mineral resources. They also compile data on known deposits, conduct research into the processes that form mineral deposits, track mineral commodity production and use and update assessments to estimate the amounts, quality and location of future resources on local, national and global scales.

In most countries, exploration and discovery is done largely by private companies and individuals, building on, but independent of, the underlying research and data development of universities and governmental geological surveys. Some companies conduct their own research or fund others to do targeted research. However, private companies are particularly susceptible to economic cycles with a clear correlation between economic cycles and exploration expenditures (Figure 5). A downsized mineral industry during a low cycle may not be able to respond quickly to the next up cycle, thus resulting in short-term mismatches in supply and demand. For the past 10 years, greatly increased exploration expenditures do not appear to have resulted in proportional discovery success, a trend some attribute to exploration for increasingly deeper ore bodies that are undercover or in remote regions. Thus, adequate supplies of mineral resources for future generations should not be taken as a given. The need for mineral resources is clear, but the path to a sustainable future will reflect the distribution of materials in the Earth and ultimately will depend on the choices, innovation, policies and values of human society. This deserves serious thought.

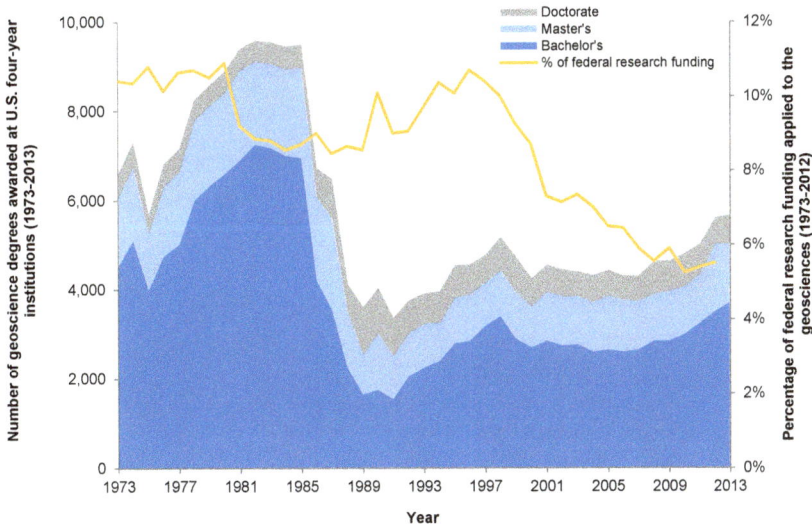

Figure 4. Geoscience degrees granted by year and U.S. Federal funding of geoscience as a percentage of total research spending. Copyright 2014 American Geosciences Institute and modified with their permission [75].

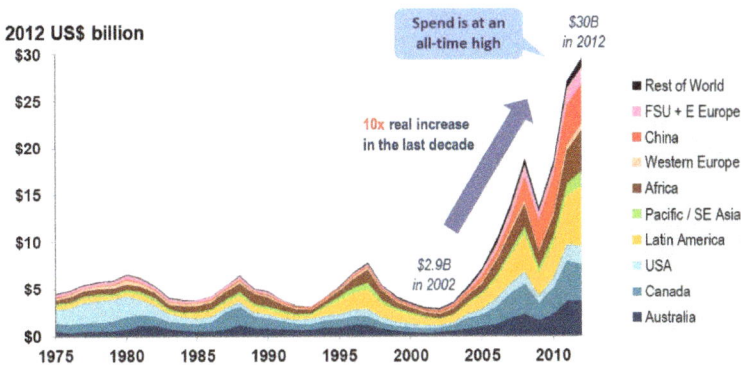

Figure 5. World exploration expenditure. Source: [76].

Even with the problems identified above, the future is not grim. Discoveries continue to be made even in well-explored districts. Most exploration has been concentrated in the upper part of the upper kilometer of the Earth's crust, even though economic deposits have been discovered down to three or more kilometers. Thus, deeper drilling combined with new exploration technology promises new discoveries, even in mature districts. We do not know what the limits to mineral resources might be, but we do know that we have not come close to reaching them yet.

Acknowledgments: We wish to acknowledge the many scientists of the USGS who have compiled over many decades the data that underlie this and most studies needing minerals information. We received thoughtful informal reviews from Nick Arndt, Dan Edelstein, Rich Goldfarb, Jane Hammarstrom, Simon Jowitt, Jon Price and Brian Skinner, as well as formal reviews by John DeYoung and Jeff Hedenquist. We thank all reviewers for their insights, but any shortcomings of the final manuscript are solely the responsibility of the authors.

Author Contributions: L. Meinert received the original invitation to write the paper and wrote the bulk of the text. G. Robinson wrote the analysis of the USGS study of undiscovered copper resources. N. Nassar wrote the sections dealing with recycling and substitution, as well as doing the analysis for and creating Figure 2.

Conflicts of Interest: The authors declare no conflict of interest.

Abbreviations

The following abbreviations are frequently used in this manuscript:

Bt	billion metric tons
CRIRSCO	Committee for Mineral Reserves International Reporting Standards
Cu	copper
GDP	gross domestic production
JORC	Australian Joint Ore Reserves Committee
Mt	million metric tons
SAMREC	South African Mineral Resource Committee
USGS	U.S. Geological Survey

References

1. Malthus, T.R. *An Essay on the Principle of Population*; J. Johnson: London, UK, 1798.
2. Meadows, D.H.; Meadows, D.L.; Randers, J.; Behrens, W.W. *The Limits to Growth*; Universe Books: New York, NY, USA, 1972.
3. Gordon, R.; Koopmans, T.C.; Nordhaus, W.D.; Skinner, B.J. *Toward a New Iron Age? Quantitative Modeling of Resource Exhaustion*; Harvard University Press: Cambridge, MA, USA, 1987.
4. Ayres, R.U. Cowboys, cornucopians and long-run sustainability. *Ecol. Econ.* **1993**, *8*, 189–207. [CrossRef]
5. Deffeyes, K.S. *Hubbert's Peak—The Impending World Oil Shortage*; Princeton University Press: Princeton, NJ, USA, 2001.
6. Deffeyes, K.S. *Beyond Oil—The View from Hubbert's Peak*; Farrar, Straus and Giroux: New York, NY, USA, 2005.
7. Gordon, R.B.; Bertram, M.; Graedel, T.E. Metal stocks and sustainability. *Proc. Natl. Acad. Sci. USA* **2006**, *103*, 1209–1214. [CrossRef] [PubMed]
8. Prior, T.; Giurco, D.; Mudd, G.; Mason, L.; Behrisch, J. Resource depletion, peak minerals and the implications for sustainable resource management. *Glob. Environ. Chang.* **2012**, *22*, 577–587. [CrossRef]
9. Northey, S.; Mohr, S.; Mudd, G.; Weng, Z.; Giurco, D. Modelling future copper ore grade decline based on a detailed assessment of copper resources and mining. *Resour. Conserv. Recycl.* **2014**, *83*, 190–201. [CrossRef]
10. Ragnarsdottir, K.V.; Sverdrup, H.U. Limits to growth revisited. *Geoscientist* **2015**, *25*, 10–16.
11. Blondel, F.; Lasky, S.G. Mineral reserves and mineral resources. *Econ. Geol.* **1956**, *51*, 686–697. [CrossRef]
12. U.S. President's Materials Policy Commission (The Paley Commission). *Resources for Freedom—Foundations for Growth and Security*; U.S. Government Printing Office: Washington, DC, USA, 1952; Volume I.
13. Hubbert, M.K. Energy Resources. In *Resources and Man*; Cloud, P., Ed.; W.H. Freeman: San Francisco, CA, USA, 1969; pp. 157–239.

14. DeYoung, J.H.J.; Singer, D.A. Physical Factors that Could Restrict Mineral Supply. In *Economic Geology 75th Anniversary Volume, 1905–1980*; Skinner, B.J., Ed.; Lancaster Press, Inc.: Lancaster, PA, USA, 1981; pp. 939–954.

15. Simon, J.L. *The Ultimate Resource 2*; Princeton University Press: Princeton, NJ, USA, 1998.

16. Lomborg, B. *The Skeptical Environmentalist*; Cambridge University Press: Cambridge, UK, 2001.

17. Beckerman, W. *A Poverty of Reason—Sustainable Development and Economic Growth*; The Independent Institute: Oakland, CA, USA, 2003.

18. Tilton, J.E. Depletion and the Long-Run Availability of Mineral Commodities. In *Wealth Creation in the Minerals Industry: Integrating Science, Business, and Education*; Special Publication, No. 12; Doggett, M.D., Parry, J.R., Eds.; Society of Economic Geologists, Inc.: Easton, MD, USA, 2006; pp. 61–70.

19. Tilton, J.E. Is Mineral Depletion a Threat to Sustainable Mining? Available online: http://inside.mines.edu/UserFiles/File/economicsBusiness/Tilton/Sustainable_Mining_Paper.pdf (accessed on 23 February 2016).

20. Strauss, S.D. *Trouble in the Third Kingdom—Minerals industry in transition*; Mining Journal Books: London, UK, 1986.

21. Sabin, P. *The Bet*; Yale University Press: New Haven, CT, USA, 2013.

22. Hewett, D.F. Cycles in Metal Production. In *Transactions of the American Institute of Mining and Metallurgical Engineers Yearbook*; American Institute of Mining and Metallurgical Engineers: Littleton, CO, USA, 1929; pp. 65–98.

23. The Economist The end of the Oil Age. Available online: http://www.economist.com/node/2155717/print (accessed on 2 February 2016).

24. Bryant, J. *3:59.4—The Quest to Break the 4 Minute Mile*; Hutchinson: London, UK, 2005.

25. Cohen, D. Earth audit. *New Sci.* **2007**, *26*, 34–41. [CrossRef]

26. Sackett, P.D. Endangered Elements—Conserving the Building Blocks of Life. In *Creating a Sustainable and Desirable Future*; Costanza, R., Kubiszewski, I., Eds.; World Scientific Publishing Co.: Singapore, 2014; pp. 239–248.

27. Joint Ore Reserves Committee (JORC). *Australasian Code for Reporting of Exploration Results, Mineral Resources and ore Reserves (The JORC Code)*; Joint Ore Reserves Committee (JORC): Carlton South, Australia, 2012.

28. South African Mineral Resource Committee (SAMREC). *South African Code for Reporting of Mineral Resources and Mineral Reserves (The SAMREC Code)*; South African Institute of Mining and Metallurgy: Johannesburg, South Africa, 2000.

29. CIM Standing Committee on Reserve Definitions. *CIM Definition Standards on Mineral Resources and Mineral Reserves*; The Canadian Institute of Mining, Metallurgy and Petroleum: Westmount, QC, Canada, 2004.

30. CRIRSCO (The Committee for Mineral Reserves International Reporting Standards). *International Reporting Template for the Public Reporting of Exploration Results, Mineral Resources and Mineral Reserves*; CRIRSCO: London, UK, 2013.

31. U.S. Bureau of Mines; U.S. Geological Survey. *Principles of a Resource/Reserve Classification for Minerals*; U.S. Geological Survey Circular 831: Reston, VA, USA, 1980.

32. United Nations Economic Commission for Europe. *Framework Classification for Fossil Energy and Mineral Resources*; United Nations Economic Commission for Europe: Geneva, Switzerland, 2003.

33. Institute of Mining Engineers of Chile (IIMCh). *Code for the Certification of Exploration Prospects, Mineral Resources and Ore Reserves*; Institute of Mining Engineers of Chile: Santiago, Chile, 2004.

34. The State Bureau of Quality and Technical Supervision. *Classification for Resources/Reserves of Solid Fuels and Mineral Commodities (National Standards GB/T17766-1999)*; Chinese Standard Publishing House: Beijing, China, 1999.

35. Joint Committee of the Venture Capital Segment of the Lima Stock Exchange. *Code for Reporting on Mineral Resources and Ore Reserves*; Bolsa de Valores de Lima (Lima Stock Exchange): Lima, Peru, 2003.

36. Diatchkov, S. Principles of classification of reserves and resources in the CIS countries. *Min. Eng.* **1994**, *46*, 214–217.

37. Jakubiak, Z.; Smakowski, T. Classification of mineral reserves in former Comecon countries. *Geol. Soc. Lond. Spec. Publ.* **1994**, *79*, 17–28. [CrossRef]

38. U.S. Securities and exchange commission. *Industry Guide 7—Description of Property by Issuers Engaged or to Be Engaged in Significant Mining Operations*; U.S. Securities and exchange commission: Washington, DC, USA, 2016.

39. SEC Reserves Working Group of SME Resources and Reserves Committee. *Recommendations concerning Estimation and Reporting of Mineral Resources and Mineral Reserves, Prepared for Submission to the United States Securities and Exchange Commission*; The Society for Mining, Metallurgy and Exploration, Inc.: Engelwood, CO, USA, 2005.

40. Doggett, M.D. Global Mineral Exploration and Production—The Impact of Technology. In *Proceedings for a Workshop on Deposit Modeling, Mineral Resource Assessment, and Their Role in Sustainable Development*; Briskey, J.A., Schultz, K.J., Eds.; U.S. Geological Survey: Reston, VA, USA, 2007; pp. 63–68.

41. Tilton, J.E. Assessing the long-run availability of copper. *Resour. Policy* **2007**, *32*, 19–23. [CrossRef]

42. Kelly, T.; Buckingham, D.; DiFrancesco, C.; Porter, K.; Goonan, T.; Sznopek, J.; Berry, C.; Crane, M. *Historical Statistics for Mineral Commodities in the United States*; Open-File Report 01–006; U.S. Geological Survey: Reston, VA, USA, 2005.

43. U.S. Geological Survey. *Mineral Commodity Summaries 2016*; U.S. Geological Survey: Reston, VA, USA, 2016.

44. Hubbert, M.K. *Energy Resources—A Report to the Committee on Natural Resources*; National Academy Press: Washington, DC, USA, 1962.

45. Kerr, R.A. Peak oil production may already be here. *Science* **2011**, *331*, 1510–1511. [CrossRef] [PubMed]

46. Harris, D.P. Conventional crude oil resources of the United States—Recent estimates, methods for estimation and policy considerations. *Mater. Soc.* **1977**, *1*, 263–286.

47. U.S. Energy Information Administration (EIA). Petroleum and Other Liquids. Available online: http://www.eia.gov/petroleum/data.cfm#crude (accessed on 23 February 2016).

48. Barnett, H.J.; Morse, C. *Scarcity and Growth—The Economics of Natural Resource Availability*; Resources for the Future: Washington, DC, USA, 1963.

49. Smith, V.K. *Scarcity and Growth—Reconsidered*; Resources for the Future: Washington, DC, USA, 1979.

50. Simpson, D.R.; Toman, M.A.; Ayres, R.U. *Scarcity and Growth Revisted*; Resources for the Future: Washington, DC, USA, 2005.

51. Nickless, E.; Bloodworth, A.; Meinert, L.; Giurco, D.; Mohr, S.; Littleboy, A. *Resourcing Future Generations White Paper—Mineral Resources and Future Supply*; International Union of Geological Sciences: Paris, France, 2014.

52. Graedel, T.E.; Harper, E.M.; Nassar, N.T.; Reck, B.K. On the materials basis of modern society. *Proc. Natl. Acad. Sci. USA* **2015**, *112*, 6295–6300. [CrossRef] [PubMed]

53. Mudd, G.M.; Weng, Z.; Jowitt, S.M. Assessment of global Cu resource trends and endowments. *Econ. Geol.* **2013**, *108*, 1163–1183. [CrossRef]

54. Johnson, K.M.; Hammarstrom, J.M.; Zientek, M.L.; Dicken, C.L. *Estimate of Undiscovered Copper Resources of the World*; U.S. Geological Survey: Reston, VA, USA, 2014.

55. Kerr, R.A. The coming copper peak. *Science* **2014**, *343*, 722–724. [CrossRef] [PubMed]

56. U.S. Geological Survey Copper Statistics. In *Historical Statistics for Mineral and Material Commodities in the United States. Data Series 140*; Kelly, T.D., Matos, G.R., Eds.; U.S. Geological Survey: Reston, VA, USA, 2014.

57. Gordon, R.B. Production residues in copper technological cycles. *Resour. Conserv. Recycl.* **2002**, *36*, 87–106. [CrossRef]

58. U.S. Census Bureau International Data Base. Available online: http://www.census.gov/population/international/data/idb/informationGateway.php (accessed on 23 February 2016).

59. Menzie, W.D.; DeYoung, J.H., Jr.; Steblez, W.G. *Some Implications of Changing Patterns of Mineral Consumption*; U.S. Geological Survey Open-File Report 03–382; U.S. Geological Survey: Reston, VA, USA, 2003.

60. Graedel, T.E.; Cao, J. Metal spectra as indicators of development. *Proc. Natl. Acad. Sci.* **2010**, *107*, 20905–20910. [CrossRef] [PubMed]

61. Graedel, T.E.; van Beers, D.; Bertram, M.; Fuse, K.; Gordon, R.B.; Gritsinin, A.; Kapur, A.; Klee, R.J.; Lifset, R.J.; Memon, L.; *et al.* Multilevel cycle of anthropogenic copper. *Environ. Sci. Technol.* **2004**, *38*, 1242–1252. [CrossRef] [PubMed]

62. Kesler, S.E. Mineral supply and demand into the 21st century. In *Proceedings for a Workshop on Deposit Modeling, Mineral Resource Assessment, and Their Role in Sustainable Development. Circular 1294*; U.S. Geological Survey: Reston, VA, USA, 2007; pp. 55–62.

63. International Copper Study Group. *ICSG 2015 Statistical Yearbook*; International Copper Study Group: Lisbon, Portugal, 2015.

64. World Bank Open Data. Available online: http://data.worldbank.org/ (accessed on 23 February 2016).

65. Graedel, T.E.; Allwood, J.; Birat, J.P.; Buchert, M.; Hagelüken, C.; Reck, B.K.; Sibley, S.F.; Sonnemann, G. What do we know about metal recycling rates? *J. Ind. Ecol.* **2011**, *15*, 355–366. [CrossRef]

66. Hein, J.R.; Mizell, K.; Koschinsky, A.; Conrad, T.A. Deep-ocean mineral deposits as a source of critical metals for high- and green-technology applications—comparison with land-based resources. *Ore Geol. Rev.* **2013**, *51*, 1–14. [CrossRef]

67. Cathles, L.M. A Path Forward. Available online: https://www.segweb.org/pdf/views/2010/08/SEG-Newsletter-Views-Lawrence-Cathles.pdf (accessed on 23 February 2016).

68. Robinson, G.R. J.; Menzie, W.D. *Economic Filters for Evaluating Porphyry Copper Deposit Resource Assessments Using Grade-Tonnage Deposit Models, with Examples from the U.S. Geological Survey Global Mineral Resource Assessment*; U.S. Geological Survey Scientific Investigations Report 2010–5090–H; U.S. Geological Survey: Reston, VA, USA, 2012.

69. Kesler, S.; Wilkinson, B.H. Earth's copper resources estimated from tectonic diffusion of porphyry copper deposits. *Geology* **2008**, *36*, 255–258. [CrossRef]

70. Gerst, M.D.; Graedel, T.E. In-Use Stocks of Metals: Status and Implications. *Environ. Sci. Technol.* **2008**, *42*, 7038–7045. [CrossRef] [PubMed]

71. Graedel, T.E.; Dubreuil, A.; Gerst, M.D.; Hashimoto, S.; Moriguchi, Y.; Müller, D.; Pena, C.; Rauch, J.; Sinkala, T.; Sonnemann, G. *Metal Stocks in Society—Scientific Synthesis*; United Nations Environment Programme: Paris, France, 2010.

72. Ciacci, L.; Reck, B.K.; Nassar, N.T.; Graedel, T.E. Lost by design. *Environ. Sci. Technol.* **2015**, *49*, 9443–9451. [CrossRef] [PubMed]

73. Nassar, N.T. Limitations to elemental substitution as exemplified by the platinum-group metals. *Green Chem.* **2015**, *17*, 2226–2235. [CrossRef]

74. Vidal, O.; Goffé, B.; Arndt, N. Metals for a low-carbon society. *Nat. Geosci.* **2013**, *6*, 894–896. [CrossRef]

75. Wilson, C. *Status of the Geoscience Workforce 2014*; American Geosciences Institute: Washington, DC, USA, 2014.

76. Schodde, R.C. *The Impact of Commodity Prices and Other Factors on the Level of Exploration*; Centre for Exploration Targeting, University of Western Australia: Perth, Australia, 2013.

resources

MDPI

Review

The Abiotic Depletion Potential: Background, Updates, and Future

Lauran van Oers [†],[*] and Jeroen Guinée [†]

Faculty of Science Institute of Environmental Sciences (CML)—Department of Industrial Ecology,
Leiden University, P.O. Box 9518, Leiden 2300, RA, The Netherlands
* Correspondence: oers@cml.leidenuniv.nl; Tel.: +31-(0)71-527-5640
† These authors contributed equally to this work.

Academic Editors: Damien Giurco and Mario Schmidt
Received: 17 December 2015; Accepted: 23 February 2016; Published: 2 March 2016

Abstract: Depletion of abiotic resources is a much disputed impact category in life cycle assessment (LCA). The reason is that the problem can be defined in different ways. Furthermore, within a specified problem definition, many choices can still be made regarding which parameters to include in the characterization model and which data to use. This article gives an overview of the problem definition and the choices that have been made when defining the abiotic depletion potentials (ADPs) for a characterization model for abiotic resource depletion in LCA. Updates of the ADPs since 2002 are also briefly discussed. Finally, some possible new developments of the impact category of abiotic resource depletion are suggested, such as redefining the depletion problem as a dilution problem. This means taking the reserves in the environment and the economy into account in the reserve parameter and using leakage from the economy, instead of extraction rate, as a dilution parameter.

Keywords: ADP; abiotic depletion potential; life cycle assessment; abiotic natural resources; elements; minerals; resource availability; scarcity; criticality; reserves

1. Introduction

From the beginning of the life cycle assessment (LCA) approach, the depletion of abiotic resources has been one of the impact categories taken into account in the environmental impact assessment. Natural resources are defined as an area of protection by the SETAC WIA (Society of Environmental Toxicology and Chemistry Working group on life cycle Impact Assessment) [1] and are part of the Life Cycle Impact Midpoint-Damage Framework developed by the UNEP (United Nations Environment Program)/SETAC life cycle initiative [2].

However, abiotic resource depletion is one of the most debated impact categories because there is no scientifically "correct" method to derive characterization factors [3]. There are several reasons for this: (1) abiotic depletion is a problem crossing the economy–environment system boundary, since reserves of resources depend on future technologies for extracting them; (2) there are different ways to define the depletion problem, and all can be justified from different perspectives; (3) there are different ways of quantifying a depletion definition, and none of them can be empirically verified, since they all depend on the assumed availability of, and demand, for resources in the future and on future technologies.

The debate on abiotic resource depletion and how to evaluate it has recently started again. This is partly because of the ongoing debate in the LCA community; see for example the guidelines of the International Reference Life Cycle Data System (ILCD) and the PEF, in Europe. (The ILCD Handbook on LCA aims to provide guidance for good practice in LCA in business and government. The development of the ILCD was coordinated by the European Commission and has been carried out in a broad international consultation process with experts, stakeholders, and the general public [4,5].

DG Environment has worked together with the European Commission's Joint Research Centre (JRC IES) and other European Commission services towards the development of a harmonized methodology for the calculation of the environmental footprint of products (including carbon). This methodology has been developed building on the ILCD Handbook as well as other existing methodological standards and guidance documents (ISO 14040-44, PAS 2050, BP X30, WRI/WBCSD GHG protocol, Sustainability Consortium, ISO 14025, Ecological Footprint, *etc.*) [6]. In addition, the debate on the criticality of resources has revived the debate on how to evaluate the use and depletion of resources by society [7–13].

In the context of the ILCD handbook on LCA, different characterization models for abiotic resource depletion have been reviewed by the LCA impact assessment community [12,14]. The characterization factors for abiotic resource depletion defined by Oers *et al.* [15] and recommended in the Dutch LCA Handbook [16], were selected as the best available operational method at present for so-called "use to availability ratio" methods [12]. However, contrary to the baseline method recommended in the Dutch LCA Handbook [16], the ILCD handbook and the PEF adopted a version of the abiotic depletion potential (ADP) that is calculated using the reserve base instead of the ultimate reserve estimations. This alternative choice was one of the reasons why the debate was resumed.

In this context it is useful to reflect on the assumptions that were made when developing the ADP and to think about possible future developments. This article aims to briefly describe the background considerations, options, and final choices made at the time of the original development (1995) and the latest update (2002) of the ADP. This description largely builds on the elaborate reporting of the original method developed by the Leiden Institute of Environmental Sciences (CML) [3,15,16].

2. Description of the Characterization Model for ADP, Considerations, Options, and Choices

2.1. Fundamentals and Choices (1995–2002)

Life cycle impact assessment (LCIA) is the phase in which the set of results of the inventory analysis—mainly the inventory table—is further processed and interpreted in terms of environmental impacts. Based on an evaluation, the different elementary flows contributing to a specific impact category are aggregated into one impact score. Thus, the core issue addressed by the characterization model for abiotic resource depletion is: how serious is the depletion of one particular natural resource in relation to that of another, and how can this be expressed in terms of characterization factors (ADPs) for these resources?

The development of the model requires many decisions to be made, which together frame the problem. This paper focuses only on the depletion problem of abiotic resource deposits [3,15]. The present section describes a selection of these issues and choices in more detail.

2.1.1. Definition of the Problem

When conducting an environmental assessment, it is debatable whether or not abiotic resource depletion should be part of the environmental impact assessment. After all, the problem mainly refers to the depletion of functions that natural resources have for the economy. One might, therefore, argue that resource depletion is basically an economic problem, rather than an environmental problem. This would imply that no separate impact category should be defined for the depletion of resources. Note that the environmental impact of the extraction process itself will, however, still be assessed through the contribution of current extraction processes to other impact categories.

Next to this, the problem of depletion of abiotic resources can still be defined in different ways, such as a decrease in the amount of the resource itself, a decrease in world reserves of useful energy/exergy, or an incremental change in the environmental impact of extraction processes at some point in the future (e.g., due to having to extract lower-grade ores or recover materials from scrap) *etc.* [12,16–18].

In Guinée and Heijungs [3] and Oers *et al.* [15], resource depletion was considered an environmental problem in its own right, while recognizing that views differ on this. The problem was defined as the decreasing natural availability of abiotic natural resources, including fossil energy resources, elements, and minerals.

2.1.2. Concepts for Assessing Depletion

How can the "decreasing availability" of a given resource be determined? In other words, what are possible indicators of resource depletion? The number of indicators that have been proposed even exceeds that of the definitions (see for an overview, for example, ILCD [12,14] and Klinglmair *et al.* [8]).

Many discussions focus on the dichotomy between price-based and physics-based indicators. Although the price of a resource can be regarded as a measure of its scarcity and societal value, it reflects more than just that. Prices are also influenced by the structure of particular economic markets, national social conditions reflected in labor cost, the power of mining companies with a monopoly, the costs of identifying new reserves, *etc.* For these reasons, prices of resources do not seem to be an appropriate indicator of depletion.

A depletion indicator could also be based on the various unique functions that resources can fulfill in materials and products. When trying to assess the availability of possible resources one would like to take into account possibilities for substitution. Oers *et al.* [15] undertook a preliminary exploration of taking substitution possibilities into account. However, elements and compounds may have very different potential functions, and possible shifts in potential functions in the future are very difficult to anticipate. Hence, it was concluded at the time that including substitution was not feasible in a characterization model for resource depletion. An exception was made for fossil energy carriers, as they were assumed to be fully interchangeable, particularly regarding their energy carrier function. It was therefore suggested to define a separate impact category for fossil fuels, based on their similar function as energy carriers [15]. However, this recommendation was not yet implemented in the baseline characterization factors described in the Dutch LCA Handbook [16].

Guinée and Heijungs [3] decided to base the characterization model for abiotic resource depletion on physical data on reserves and annual de-accumulation, with de-accumulation defined as the annual production (e.g., in kg/yr) minus the annual regeneration (e.g., in kg/yr) of a resource, the latter of which was assumed to be zero. In addition to this, Oers *et al.* decided that the implementation of substitution options (which touches upon issues of scarcity and criticality) was not (or not yet) feasible within LCA [15].

2.1.3. Definition of Availability and Natural Stocks *Versus* Stocks in the Economy

When assessing the availability of resources one can use the concept of availability in a narrow or a broad sense. Availability in the narrow sense focuses on the extraction of the resource from the stock in the environment, the primary extraction medium, whereas availability in the broad sense focuses on the presence of resources in stocks in the environment as well as the economy (geo- and anthropospheres).

Ideally based on the definition of the depletion of abiotic resources, the available resource should encompass both natural stocks and stocks in the economy. The criterion for depletion of the resource is whether the resource derived from the environment is still present and (easily) available in the stocks of materials in the economy. After all, as long as resources are still available in the economic stock after extraction, there is no depletion problem.

Guinée and Heijungs [3] and Oers *et al.* [15] decided to adopt the narrow definition of availability, while recognizing that, eventually, a broad sense definition would be preferable, assuming that it would be possible and practically feasible to define a proper indicator for this and that the necessary data would be available. The Discussion section below briefly introduces a preview of a possible new approach.

2.1.4. Types of Reserves and Definitions

Estimates of the amounts of resources (elements, minerals, fuels) available for future generations depend on the definition of reserve that is used. When talking about the reserves of resources there might be confusion about the type of reserve being considered. The LCIA and geological community do not use the same definitions as traditionally used by leading geological institutions. Drielsma *et al.* [18] have compared the definitions as used by the Committee for Mineral Reserves International Reporting Standards (CRIRSCO) with definitions of reserves as used in the ADP [15] (Table 1). For better communication between both communities in the future the terminology of resources and reserves should be harmonized. Within the geological community the institutions are currently converging towards the CRIRSCO definitions. It seems logical that within the LCIA community the same terminology and definitions will be adopted.

Table 1. Types of reserves and definitions.

Terminology		Definition
Oers *et al.* [15]	Drielsma *et al.* [19]	A Resource/Reserve Classification for Minerals, USGS [3,20,21].
ultimate reserve	crustal content	The quantity of a resource (like a chemical element or compound) that is ultimately available, estimated by multiplying the average natural concentration of the resource in the primary extraction media (e.g., the earth's crust) by the mass or volume of these media (e.g., the mass of the crust assuming a depth of e.g., 10 km) [3].
ultimately extractable reserve	extractable global resource	Those reserves that can ultimately be technically extracted may be termed the "ultimately extractable reserves". This ultimately extractable reserve ("extractable global resource") is situated somewhere between the ultimate reserve and the reserve base [20,21].
reserve base	mineral resource	Part of an identified resource that meets specified minimum physical and chemical criteria relating to current mining practice. The reserve base may encompass those parts of the resources that have a reasonable potential for becoming economically available within planning horizons beyond those that assume proven technology and current economics. The reserve base includes those resources that are currently economic (reserves) or marginally economic (marginal reserves), and some of those that are currently subeconomic (subeconomic resources) (for further definitions see the original references) [20,21].
economic reserve	mineral reserve	The part of the natural reserve base which can be economically extracted at the time of determination [20,21].

The disadvantage of the "reserve base" and "economic reserve" is that estimating the size of the reserve involves a variety of technical and economic considerations not directly related to the environmental problem of resource depletion. The estimates, however, are relatively certain, as they are based on present practice while, on the other hand, they are highly unstable as they continuously change over time. In contrast, the "ultimately extractable reserve" is more directly related to the environmental problem of resource depletion, and relatively stable over time. However, it is highly uncertain how much of the scattered concentrations of elements and compounds will eventually become available, as technical and economic developments in the far future are unpredictable.

The ultimate reserve and ultimately extractable reserve are expected to differ substantially. However, data on the ultimately extractable reserve are unavailable and will never be exactly known because of their dependence on future technological developments. Nevertheless, one might assume that the "ultimate reserve" is a proxy for the "ultimately extractable reserve", implicitly assuming that the ratio between the ultimately extractable reserve and the ultimate reserve is equal for all resource types. In reality this will not be the case, because the concentration-presence-distribution (see Figure 1) of different resources will most likely be different [15]. Hence, there is insufficient information to decide which of these reserves gives the best indication of the ultimately extractable reserve. Whilst we acknowledge some authors propose a mineralogical barrier as described by Skinner [22] this has not been considered in the research.

Guinée and Heijungs [3] and Oers *et al.* [15] adopted the "ultimate reserve" as the presumably best proxy for the "ultimately extractable reserve" in their characterization model for abiotic resource depletion. They recommended that alternative indicators be used for a sensitivity analysis, like the "reserve base", and to a lesser extent the "economic reserve".

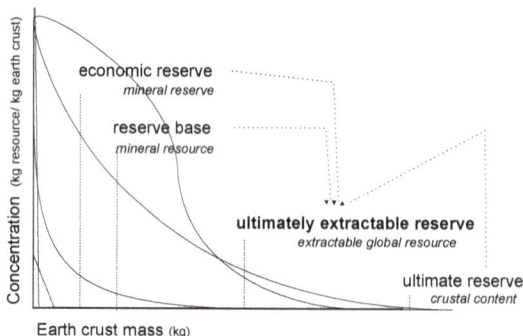

Figure 1. Concentration-presence-distribution of several theoretical resources in the Earth's crust. The average Earth crust thickness is assumed to be 17 km. The Earth crust surface is assumed to be 5.14×10^{14} m^2. The average earth crust density is 2670 kg·m^{-3}. The ultimate reserve of a resource is the surface area enclosed by the curve. The size of the other estimates of the reserves is given by the surface area enclosed by the curve and the given secant with the x-axis [15].

2.1.5. Equations for Characterization Factors

Based on all the choices described above, the characterization model can be described. The haracterization model is a function of natural reserves (stocks/deposits in the environment) of the abiotic resources combined with their rates of extraction (see Equation (2)). The method has been made operational for many elements and fossil fuels (actually: the energy content of fossil fuels). The natural reserves of these resources are based on "ultimate reserves"; that is, on concentrations of the elements and fossil carbon in the Earth's crust.

The characterization factor is the abiotic depletion potential (ADP). This factor is derived for each extraction of elements and fossil fuels and is a relative measure, with the depletion of the element antimony as a reference (see Equation (2)).

In accordance with the general structure of the LCIA, the impact category indicator result for the impact category of "abiotic depletion" is calculated by multiplying LCI results, extractions of elements and fossil fuels (in kg) by the characterization factors (ADPs in kg antimony equivalents/kg extraction, The choice of the reference substance is arbitrary. Choosing another reference will not change the relative sizes of the characterization factors. Antimony was chosen as a reference substance because it is the first element in the alphabet for which a complete set of necessary data (extraction rate and ultimate reserve) is available, and aggregating the results of these multiplications in one score to obtain the indicator result (in kg antimony equivalents) (see Equation (1)):

$$abiotic\ depletion = \sum_i ADP_i \times m_i \tag{1}$$

with:

$$ADP_i = \frac{DR_i/(R_i)^2}{DR_{ref}/\left(R_{ref}\right)^2} \tag{2}$$

where,

ADP_i: abiotic depletion potential of resource i (kg antimony equivalents/kg of resource i);
m_i: quantity of resource i extracted (kg);
R_i: ultimate reserve of resource i (kg);
DR_i: extraction rate of resource i (kg·yr^{-1}) (regeneration is assumed to be zero);
R_{ref}: ultimate reserve of the reference resource, antimony (kg);
DR_{ref}: extraction rate of the reference resource, Rref (kg·yr^{-1}).

The first operational set of ADPs was developed by Guinée [3]. In 2002, these ADPs were updated by Oers *et al.* [15]. This update included new extraction rates (DRs) of resources for the base year 1999. To facilitate sensitivity analysis, alternative ADPs were developed based on different definitions of reserves, *viz.* economic reserve, reserve base and ultimate reserve. Oers *et al.* [15] adjusted the ADPs of fossil fuels and defined a separate impact category for fossil fuels based on the assumption that fossil fuels are mutually substitutable as energy carriers. However, in the LCA handbook [16], fossil fuels and elements were still considered to be part of one impact category, "abiotic resource depletion". The split into two separate impact categories was implemented in 2009.

2.2. Developments after 2002

Since 2002, the ADPs have been reported in a spreadsheet together with characterization factors for other impact categories. The Centrum voor Milieuwetenschappen Leiden Impact Assessment (CMLIA) spreadsheet with impact assessment factors as recommended by the Dutch LCA Handbook [16] can be downloaded from the CMLIA website [23].

2.2.1. Update of Impact Categories by CML: Impact Category of "Abiotic Resource Depletion" Split into Two Separate Impact Categories

In 2009, the impact category of "abiotic resource depletion" was split into two separate impact categories:

- "abiotic resource depletion—elements"; and
- "abiotic resource depletion—fossil fuels".

The impact category for elements is a heterogeneous group, consisting of elements and compounds with a variety of functions (all functions being considered of equal importance). The resources in the impact category of fossil fuels are fuels like oil, natural gas, and coal, which are all energy carriers and assumed to be mutually substitutable. As a consequence, the stock of the fossil fuels is formed by the total amount of fossil fuels, expressed in Megajoules (MJ).

Despite the fact that uranium is also an energy carrier, the extraction of uranium is classified under the impact category of "abiotic resource depletion—elements" and not together with the fossil fuels, under the impact category of "abiotic resource depletion—energy carriers". Uranium and fossil fuels also have other applications besides "energy carrier". Fossil fuels are considered to be interchangeable for these other applications, like the production of plastics, while uranium is not. However, one might argue that the largest application of both fossil fuels and uranium is that of an energy carrier, and for this reason they should be classified under the same impact category. Future versions of the CMLIA might be adapted accordingly.

2.2.2. Update of ADP Values by CML

In 2009, the ADPs reported in the Dutch Handbook on LCA [16], based on Guinée [3], were updated using the ADPs reported by Oers *et al.* [15] (R and DR data based on 1999 data). Since 2009, the basic data underlying the ADPs and, thus, the ADPs themselves, have not been updated anymore.

2.2.3. Update of R and DR Values by Others

Some new DR and R data have been reported by Frischknecht *et al.* [24]. However, ADPs derived from these data have not been implemented in the CMLIA spreadsheet.

Until 2010, the US Geological Survey (USGS) [20] reported the extraction rates, (economic) reserve, and reserve base data for many resources on an annual basis. However, the reserve base data have no longer been reported since 2010.

3. Discussion and Possible New Approaches in Abiotic Resource Depletion in CMLIA

3.1. Depletion, Scarcity, and Criticality

When trying to assess the "sustainability" of the use of resources by society, different definitions of the problem of the resource use are used in terms of depletion, scarcity, and criticality.

Depletion of a resource means that its presence on Earth is reduced. This refers to geological (or natural) stocks. Scarcity of a resource means that the amount available for use is, or will soon be, insufficient ("demand higher than supply flow"). Criticality of a resource means that it is scarce and, at the same time, essential for today's society. In addition to environmental aspects, criticality assessment often also considers economic, social, and geopolitical issues [9].

Interest in natural resources has recently increased, due to the growing interest in policies to ensure the security of supply of critical metals like tantalum, indium, neodymium, *etc.* [10,11]. It should be noted, however, that this assessment of the criticality of metals is often based on more criteria than environmental issues alone. The criticality of a metal is based on a mixture of environmental, geopolitical, social, and economic considerations. Furthermore, the criticality debate does not necessarily take into account the cradle-to-grave chain perspective; it often only focuses on a supply of elements at the company or national economy level.

It is our opinion that the environmental impact assessment of LCA should not strive to take into account all these different aspects of criticality assessment, *i.e.*, environment, economy, and social aspects. They may be part of a broader life cycle sustainability assessment (LCSA), but even then a general approach will be difficult, as many of the criticality aspects are highly time- and region-dependent and even differ for different stakeholders. LCSA is a framework for (cradle-to grave) system analysis which, besides environmental aspects, may also focus on economic and social issues [25,26]. However, the impact categories for "abiotic resource depletion" only deal with the environmental pillar of the sustainability assessment, and the indicator is based on the problem of depletion only.

3.2. Ultimate Reserve, Reserve Base, and Economic Reserve

The ADPs developed by Guinée [3] and Oers *et al.* [15] are based on ultimate reserves as calculated from the average element concentrations in the Earth's crust [27] assuming a crust mass of 2.31×10^{22} kg (average depth of the earth's crust (m): 17,000; average density (kg/m^3): 2670; surface of the Earth (m^2): 5.1×10^{14}) [3].

Drielsma *et al.* [19] stated that "crustal content" (a synonym of ultimate reserve) is a stable, comprehensive dataset that can be used to derive a physical estimate of resource depletion for abiotic resources. They base this conclusion on a study by Rudnick and Gao [28], which compared several studies done since the initial study by Clarke and Washington [29] on estimates for the total stock of resources. The study by Rudnick and Gao [28] provided updated figures on crustal content that can be used for updated ADPs based on ultimate reserves.

Given the fact that estimates of economic reserves are far less stable due to technological changes and economic developments, and that reserve base data are no longer provided by USGS [20], the use of "ultimate reserves" as a basis for the calculation of ADPs seems justified and confirmed.

We, therefore, argued here that the ILCD-recommended characterization model for abiotic resource depletion [12] based on Oers *et al.* [15] should be readjusted to the original reserve definition, involving ultimate reserve, instead of reserve base.

3.3. Availability in the Broad Sense, ADP Based on Stocks in Environment and Economy

Resources are not only produced from natural resource supplies, but are also recycled from the growing amount of stocks in the economy and wastes ("urban stocks"). Should these stocks of materials in the economy, which can potentially be recycled, be included when estimating the total amount of available resources? In other words, should the reserve be based on the natural reserve only or also on the reserve in the economy?

As discussed above, resources fulfill specific unique functions in materials and products. Theoretically, there is no depletion problem as long as the resource after extraction is still available in the economic stock. As a consequence, resource availability should encompass both natural stocks and stocks in the economy [15].

Schneider *et al.* [30] made preliminary attempts to derive ADPs based on stocks in both the economy and the environment, the so-called anthropogenic stock-extended abiotic depletion potential (AADP). Their first results emphasize the relevance of anthropogenic stocks for the assessment of abiotic resource depletion. However, a larger set of characterization factors and further research are needed to verify the applicability of the concept within LCA practice. [30]. The AADPs were updated in 2015, including an approach to the estimation of the ultimately extractable reserves from the environment. The update resulted in 35 operational AADPs [31]. However, gathering information on anthropogenic stocks still remains a challenge.

The problem could therefore be redefined as follows: abiotic resource depletion is the decrease in the availability of resources, both in the environment and the economy.

3.4. Emissions from Economic Stocks and Processes as an Indicator of Dilution

If we adopt the broader definition of reserve described above, the de-accumulation parameter "DR" in the ADP equation becomes meaningless. After all, the extraction (DR) of resources from environmental sources for use in the economy is just a shift from environmental stock to economic stock. Thus, when redefining the reserve parameter (R) in the ADP model, the other parameter in the model, the "extraction rate" (DR) should also be redefined.

As originally suggested by Oers *et al* [15], the emission of resources to the environment, instead of the extraction rate, might be a promising parameter for use in the characterization model. Oers *et al.* [15] suggested that the loss of resources from economic processes and stocks due to emissions of elements and compounds to air, water, and soil can be used as a measure of the dilution of resources. The argumentation is that when environmental and economic stocks are regarded as a total reserve of resources, the problem of depletion is in fact a dilution problem (see Figure 2). The assumption is that the resources that are emitted to the environment are too diffuse to be ultimately extractable and, thus, are not part of the ultimately extractable reserve. Hence, abiotic resource depletion can also be redefined as a problem of dilution of resources; that is, a reduction of concentrated reserves of resources.

Frischknecht (in: Vadenbo *et al.* [7]) also suggested focusing on the role of borrowing and dissipative resource use in impact assessment of abiotic resources. Material resources on Earth cannot be lost (unless converted into energy or lost into space) but might be dispersed. The impact assessment factors derived by Frischknecht are still based on the original ADP model (using environmental reserves only), but extraction rates are applied to the dissipative use of resources, which is defined as the difference between the amounts of resources extracted and recycled, *i.e.*, the aggregated amount lost during manufacture, use, and end-of-life treatment. This new concept of assessing the dissipative use of resources using ADP factors is recommended by the new version of the ecoscarcity method [24].

Oers *et al.* [15] suggested a different elaboration of the characterization model for the dilution problem of resources. If depletion of abiotic resources is defined as the dilution of the resources, the leakage (L) of elements, minerals, and energy (heat) from the economy is suggested as a parameter for the characterization model, as an alternative to the extraction rate (DR). This leakage of resources can then be combined with the total reserve of resources in the environment and the economy (Rtotal = Renv + Rec) into a new characterization model for the dilution of abiotic resources. This new model can still use the original ADP equation, but the parameter extraction rate (DR) is replaced by Leakage (L), and the reserve parameter (R) refers to reserves in both the economy and the environment. Making these new characterization factors operational requires an estimation of stocks of elements in the economy and the total emissions of elements to the environment in the world. A preliminary approach to estimating these stocks can be based on the method and data described by Schneider [30,31] and the United Nations Environment Program (UNEP) (stocks per capita) [9]. The total emissions in the world can be based on the inventories made to derive normalization factors, as in the work by Wegener Sleeswijk *et al.* [32].

The result for the impact category of "abiotic resource dilution" can then be calculated by multiplying the emission, instead of extraction, of elements (in kg) by the characterization factors (ADPs in kg antimony equivalents/kg emission) and by aggregating the results of these multiplications in one score to obtain the indicator result (in kg antimony equivalents).

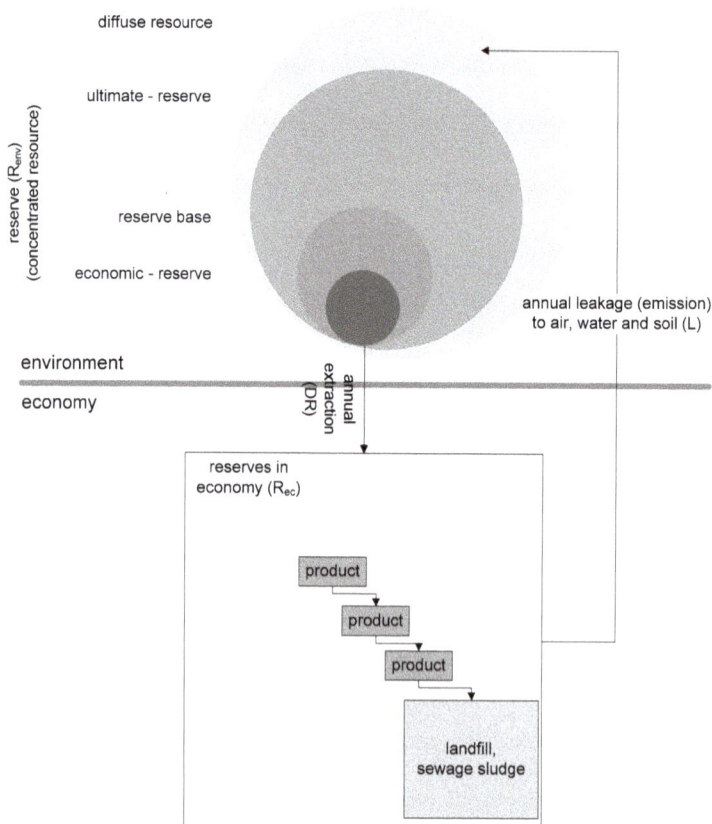

Figure 2. Relevant parameters for the abiotic resource depletion model, reserves in economy and environment and annual extraction or emission of the resource (adapted from [15]).

4. Conclusions

It is impossible to define one correct method for assessing the problem of depletion of abiotic resources, since the choice of relevant parameters that make up the model will depend on the problem definition, and the correctness of the parameters cannot be verified empirically.

However, the definition of the problem and the choices made to define the characterization model will result in different sets of characterizations.

The model of abiotic resource depletion as defined in the ADP [3,15] is a function of the annual extraction rate and geological reserve of a resource.

In the model as presently defined, the ultimate reserve is considered the best estimate of the ultimately extractable reserve and also the most stable parameter for the reserve parameter. However, data for this parameter will by definition never be available. As a proxy, we suggest the ultimate reserve (crustal content).

Extraction rates and reserves are regularly updated by USGS [19], partly as a result of changes in technology and new insights. As a consequence, characterization factors should also be regularly updated, but this is unfortunately no longer being done.

The impact category of abiotic resource depletion as defined by CML [3,15] comprises only the depletion of environmental resources. Criticality of resources is not part of the problem definition. We recommend not including criticality in environmental LCA, as it deals with more than just environmental aspects. It could be included in the broader LCSA framework, which also tries to incorporate economic and social assessment into life cycle thinking.

A possible new development for the characterization model defined by Oers *et al.* [15] is the redefinition of resource depletion as a dilution problem. This implies the inclusion of reserves in the economy into the reserve parameter and using leakage from the economy, instead of extraction rate, as a dilution parameter. However, this idea remains to be worked out, and no operational characterization factors are as yet available.

Acknowledgments: Part of the development of the ADP is based on the project "Abiotic resource depletion in LCA; Improving characterization factors for abiotic resource depletion as recommended in the new Dutch LCA handbook". This project was commissioned by the Dutch Ministry of Infrastructure and the Environment.

Conflicts of Interest: The authors declare no conflict of interest. The sponsors had no role in the design of the study; in the collection, analysis, or interpretation of data; in the writing of the manuscript; or in the decision to publish the results.

Abbreviations

The following abbreviations are used in this manuscript:

AADP	anthropogenic stock extended abiotic depletion potential
ADP	abiotic depletion potential
CML	Centrum voor Milieuwetenschappen Universiteit Leiden (Institute of Environmental Sciences, Leiden University)
CMLIA	Centrum voor Milieuwetenschappen Universiteit Leiden Impact Assessment
CRIRSCO	Committee for Mineral Reserves International Reporting Standards
DR	extraction rate of resource, originally defined as annual de-accumulation, with de-accumulation defined as the annual extraction (e.g., in kg/yr) minus the annual regeneration (e.g., in kg/yr) of a resource, the latter of which is assumed to be zero
ILCD	International Reference Life Cycle Data System
LCA	life cycle assessment
LCIA	life cycle impact assessment
LCSA	life cycle sustainability assessment
PEF	product environmental footprint
R	reserve of resource
SETAC	Society of Environmental Toxicology and Chemistry
UNEP	United Nations Environment Program
USGS	United States Geological Survey
WIA	Working Group on Life Cycle Impact Assessment

References

1. Udo de Haes, H.A., Jolliet, O., Finnveden, G., Hauschild, M., Krewit, W., Müller-Wenk, R. (Eds.) Best available practice regarding impact categories and category indicators in life cycle impact assessment. *Int. J. Life Cycle Assess.* **1999**, *4*, 167–174.
2. Jolliet, O.; Muller-Wenk, R.; Bare, J.; Brent, A.; Goedkoop, M.; Heijungs, R.; Itsubo, N.; Pena, C.; Pennington, D.; Potting, J.; *et al.* The LCIA midpoint-damage framework of the UNEP/SETAC life cycle initiative. *Int. J. Life Cycle Assess.* **2004**, *9*, 394–404. [CrossRef]
3. Guinée, J.; Heijungs, R. A proposal for the definition of resource equivalency factors for use in product Life-Cycle Assessment. *Environ. Toxicol. Chem.* **1995**, *14*, 917–925. [CrossRef]
4. European Commission. *ILCD Handbook, General Guide for Life Cycle Assessment—Detailed Guidance*; European Commission: Ispra, Italy, 2010. Available online: http://publications.jrc.ec.europa.eu/repository/bitstream/JRC48157/ilcd_handbook-general_guide_for_lca-detailed_guidance_12march2010_isbn_fin.pdf (accessed on 29 February 2016).
5. European Commission. European Platform on Life Cycle Assessment. 2015. Available online: http://eplca.jrc.ec.europa.eu/?page_id=86 (accessed on 29 February 2016).
6. European Commission. Environmental Footprint News. 2016. Available online: http://ec.europa.eu/environment/eussd/smgp/ef_news.htm (accessed on 29 February 2016).
7. Vadenbo, C.; Rørbech, J.; Haupt, M.; Frischknecht, R. Abiotic resources: New impact assessment approaches in view of resource efficiency and resource criticality—55th Discussion Forum on Life Cycle Assessment, Zurich, Switzerland, April 11, 2014. *Int. J. Life Cycle Assess.* **2014**, *19*, 1686–1692. [CrossRef]
8. Klinglmair, M; Sala, S.; Brandão, M. Assessing resource depletion in LCA: A review of methods and methodological issues. *Int. J. Life Cycle Assess.* **2014**, *19*, 580–592.
9. Van der Voet, E. Criticality and abiotic resource depletion in life cycle assessment. In *Security of Supply and Scarcity of Raw Materials. Towards a Methodological Framework for Sustainability Assessment*; Mancini, L., De Camillis, C., Pennington, D., Eds.; European Commission: Luxemburg, 2013; pp. 21–23.
10. United Nations Environment Programme. Environmental Risks and Challenges of Anthropogenic Metals Flows and Cycles. 2013. Available online: http://www.unep.org/resourcepanel/Publications/EnvironmentalChallengesMetals/tabid/106142/Default.aspx (accessed on 29 February 2016).
11. European Commission. *Critical raw materials for the EU*; Report of the Ad-hoc Working Group on defining critical raw materials; European Commission: Brussels, Belgium, 2010. Available online: http://ec.europa.eu/growth/sectors/raw-materials/specific-interest/critical/index_en.htm (accessed on 29 February 2016).
12. European Commission. In *ILCD handbook: Recommendations for Life Cycle Impact Assessment in the European context—Based on Existing Environmental Impact Assessment Models and Factors*; European Commission, Joint Research Centre, Institute for Environment and Sustainability: Ispra, Italy, 2011. Available online: http://publications.jrc.ec.europa.eu/repository/handle/JRC61049 (accessed on 29 February 2016).
13. European Commission. *Report on Critical Raw Materials for the EU*; European Commission: Brussels, Belgium, 2014. Available online: http://ec.europa.eu/growth/sectors/raw-materials/specific-interest/critical/index_en.htm (accessed on 29 February 2016).
14. European Commission. *Analysis of existing Environmental Impact Assessment methodologies for use in Life Cycle Assessment*; European Commission, Joint Research Centre, Institute for Environment and Sustainability, 2010. Available online: http://eplca.jrc.ec.europa.eu/uploads/ILCD-Handbook-LCIA-Background-analysis-online-12March2010.pdf (accessed on 29 February 2016).
15. Van Oers, L.; De Koning, A.; Guinée, J.B.; Huppes, G. Abiotic resource depletion in LCA. Improving characterisation factors for abiotic resource depletion as recommended in the new Dutch LCA handbook. RWS-DWW: Delft, The Netherlands, 2002. Available online: http://www.leidenuniv.nl/cml/ssp/projects/lca2/report_abiotic_depletion_web.pdf (accessed on 29 February 2016).
16. Guinée, J.B.; Gorée, M.; Heijungs, R.; Huppes, G.; Kleijn, R.; de Koning, A.; van Oers, L.; Wegener Sleeswijk, A.; Suh, S.; Udo de Haes, H.A.; *et al.* Handbook on Life Cycle Assessment: Operational Guide to the ISO Standards; Kluwer Academic Publisher: Dordrecht, The Netherlands, 2002. Available Online: http://www.leidenuniv.nl/cml/lca2/index.html (accessed on 29 February 2016).

17. Finnveden, G. "Resources" and related impact categories. In *Towards a Methodology for Life Cycle Impact Assessment*; Udo de Haes, H.A., Ed.; SETAC-Europe: Brussels, Belgium, 1996.
18. Heijungs, R.; Guinée, J.; Huppes, G. *Impact Categories for Natural Resources and Land Use*; CML-report 138; Leiden University: Leiden, The Netherlands, 1997.
19. Drielsma, J.A.; Russell-Vaccari, A.J.; Drnek, T.; Brady, T.; Weihed, P.; Mistry, M.; Perez Simbor, L. Mineral resources in life cycle impact assessment—Defining the path forward. *Int. J. Life Cycle Assess.* **2016**, *21*, 85–105. [CrossRef]
20. United States Geological Survey. Commodity Statistics and Information, Statistics and information on the worldwide supply of, demand for, and flow of minerals and materials essential to the U.S. economy, the national security, and protection of the environment. 2015. Available online: http://minerals.usgs.gov/minerals/pubs/commodity/ (accessed on 29 February 2016).
21. United States Geological Survey. Appendix of Commodity Statistics and Information, Statistics and information on the worldwide supply of, demand for, and flow of minerals and materials essential to the U.S. economy, the national security, and protection of the environment A Resource/Reserve Classification for Minerals. 2015. Available online: http://minerals.usgs.gov/minerals/pubs/mcs/2007/mcsapp07.pdf (accessed on 29 February 2016).
22. Skinner, B.J. Exploring the resource base. In Proceedings of the workshop on "The Long-Run Availability of Minerals", Washington, DC, USA, 22–23 April 2001; Available online: http://www.rff.org/files/sharepoint/WorkImages/Download/RFF-Event-April01-keynote.pdf (accessed on 29 February 2016).
23. CML-IA Characterisation Factors. Available online: https://www.universiteitleiden.nl/en/research/research-output/science/cml-ia-characterisation-factors (accessed on 29 February 2016).
24. Frischknecht, R.; Büsser Knöpfel, S. Swiss Eco-Factors 2013 according to the Ecological Scarcity Method. Methodological fundamentals and their application in Switzerland. Environmental studies No. 1330. Federal Office for the Environment: Bern, Switzerland, 2013; p. 254. Available online: http://www.bafu.admin.ch/publikationen/publikation/01750/index.html?lang=en (accessed on 29 February 2016).
25. UNEP-SETAC. *Towards a Life Cycle Sustainability Assessment. Making informed choices on products*. 2011. Available online: http://www.unep.org/pdf/UNEP_LifecycleInit_Dec_FINAL.pdf (accessed on 29 February 2016).
26. UNEP-SETAC. *Life Cycle Sustainability Assessment*. 2015. Available online: http://www.lifecycleinitiative.org/starting-life-cycle-thinking/life-cycle-approaches/life-cycle-sustainability-assessment/ (accessed on 29 February 2016).
27. Lide, D.R., Ed.; *CRC Handbook of Chemistry and Physics*, 71st ed.; CRC Press: Boston, MA, USA, 1990.
28. Rudnick, R.L.; Gao, S. Composition of the continental crust. In *The Crust*; Rudnick, R.L., Ed.; Elsevier: Philadelphia, PA, USA, 2005; Volume 3, pp. 1–64.
29. Clarke, F.W.; Washington, H.S. *The Composition of the Earth's Crust*; USGS Professional Paper 127; USGS: Washington, DC, USA, 1924; p. 117.
30. Schneider, L.; Berger, M.; Finkbeiner, M. The anthropogenic stock extended abiotic depletion potential (AADP) as a new parameterisation to model the depletion of abiotic resources. *Int. J. Life Cycle Assess.* **2011**, *16*, 929–936. [CrossRef]
31. Schneider, L.; Berger, M.; Finkbeiner, M. Abiotic resource depletion in LCA-background and update of the anthropogenic stock extended abiotic depletion potential (AADP) model. *Int. J. Life Cycle Assess.* **2015**, *20*, 709–721. [CrossRef]
32. Wegener Sleeswijk, A.; van Oers, L.F.C.M.; Guinée, J.B.; Struijs, J.; Huijbregts, M.A.J. Normalisation in product life cycle assessment: An LCA of the global and European economic systems in the year 2000. *Sci. Total Environ.* **2008**, *390*, 227–240. [CrossRef] [PubMed]

resources

MDPI

Article

Physical Assessment of the Mineral Capital of a Nation: The Case of an Importing and an Exporting Country

Guiomar Calvo [1,*], Alicia Valero [1], Luis Gabriel Carmona [2,3] and Kai Whiting [3]

[1]   Research Centre for Energy Resources and Consumption (CIRCE), Mariano Esquillor n.15, Zaragoza 50018, Spain; aliciavd@unizar.es
[2]   Faculty of Environmental Sciences, Universidad Piloto de Colombia. Carrera 9 No. 45A-44, Bogotá 110231, Colombia; lugacapa@gmail.com
[3]   MARETEC, Department of Mechanical Engineering, Instituto Superior Técnico, Universidade de Lisboa, Avenida Rovisco Pais, 1, Lisboa 1049-001, Portugal; whitingke@yahoo.co.uk
*   Author to whom correspondence should be addressed; gcalvose@unizar.es; Tel.: +34-876-555-624; Fax: +34-976-732-078.

Academic Editor: Mario Schmidt
Received: 23 September 2015; Accepted: 10 November 2015; Published: 16 November 2015

Abstract: Intensified mineral consumption and reserve depletion means that it is becoming increasingly important for policymakers to account for and manage national mineral capital. Exergy replacement costs (ERC), an indicator based on the second law of thermodynamics, provides a physical value of mineral loss. When only a unit mass analysis is used, the role of scarcer minerals, such as gold, is obscured. ERC can identify those minerals which are most critical and more difficult to re-concentrate. This paper compares the mineral depletion of that of Colombia and Spain for 2011, both in mass and ERC terms. The Colombian mineral balance for that year is predominately based on fossil fuel extraction and exports, whilst Spain produced industrial minerals but relied heavily upon metals and fossil fuel imports. Using exergy replacement costs, an economic analysis was carried out to determine the impact of mineral extraction, in monetary terms, should the cost of re-concentrating such minerals be taken into account. In 2011, the GDP derived from the extractive sectors of either country did not compensate the mineral resource loss, meaning that mineral patrimony is not being properly evaluated.

Keywords: exergy analysis; exergy replacement costs; domestic material consumption; assessment of mineral trade; foreign dependency

---

1. Introduction

Mineral resource depletion and mineral capital management are critical issues that need to be addressed objectively and efficiently on an international scale across disciplines and professions. According to Krausmann *et al.* [1] the global total of material extraction has multiplied eightfold since the beginning of the 20th century. The highest increase corresponds to construction minerals and ore/industrial minerals, which grew by a factor of 34 and 27, respectively. In the European Union, domestic material consumption (DMC) is the main flow indicator within material flow accounting (MFA), a system which quantifies extractive activities in tonnes [2]. DMC measures the annual amount of raw materials extracted nationally, plus imports minus exports. A related concept is domestic extraction which considers the annual amount of raw materials (except water and air), extracted on a national level. The ratio between domestic extraction (DE) and DMC may be used to indicate national dependence on mineral extraction and trade, hence why Weisz *et al.* [3] refer to it as a "domestic resource dependency" ratio. One way to complement this ratio is through exergy replacement costs (ERC) [4].

ERC quantitatively evaluate the effort, or useful energy, needed to re-concentrate extracted mineral wealth with current best available technology. ERC depend on the mineral's composition, a deposit's average ore grade and the energy intensity of the mining and beneficiation process. Consequently, scarcer minerals, such as gold or mercury, carry more weight in this non-conventional accounting process, than the common minerals of, say, limestone or phosphate rock. This is useful because mass, the predominate measure of resource extraction, and used in the DMC, does not take quality into consideration and is thus not robust enough to properly assess the loss of mineral capital. The quality of a mineral deposit is a key consideration when it comes to evaluating sustainable development, specifically in terms of mineral scarcity and criticality. Quality and not just quantity measurements, translate into a more reliable set of results from which national and international policymakers can improve sustainable assessments and make informed decisions. These themes are currently under discussion and of foremost importance in various States [5–7].

2. Methodology

Using a thermoeconomic approach, via the unit exergy and more specifically ERC, the authors undertake a comparative case study between an exporting (Colombia) and importing (Spain) country to assess the effect trade has on their respective mineral capital. Exergy, and specifically, its application in the novel thermoeconomic branch of Physical Geonomics, first introduced by Valero [8], is one method that can, through its objective measure of quality, permit a more rigorous data analysis of mineral depletion. Exergy has been traditionally used to quantitatively measure any energy, such as heat, in terms of its pure forms (mechanical work or electricity) and may be used as an efficiency gauge which states the maximum amount of work that can be obtained when a system is brought to equilibrium with the surrounding environment. This is because, from a physical perspective, and according to the Second Law of Thermodynamics, energy is always conserved at the expense of exergy, which is always destroyed, unless the process is 100 percent reversible. In the same way, exergy can be used to assess the quality of material flows such as mineral resources [9,10].

The mineral capital assessment is accomplished through ERC, on the basis that the greater the difference between the concentration of a randomly dispersed mineral in the Earth's crust (x_c) to that in an accumulation that could be mined (x_m), the greater the exergy registered. This in turn explains why the exergy (and energy) consumed in mining increases exponentially, tending towards infinity, as concentration and particle size decreases. The key to making an exergy analysis suitable for any abiotic resource evaluation—that of minerals as well as water—is the appropriateness of the chosen reference point. Reference baselines that could be used include the conventional Reference Environment [11] or the commercially dead planet, Thanatia [12]. The former is commonly used to assess the chemical exergy of elements. However, it is rejected in this instance because it does not consider the exergy required to recuperate minerals in terms of their crustal concentration. The baseline reference employed in this paper is that of Thanatia because it can be used to give an estimate of the Earth's current level of degradation, since it includes a list of minerals with their respective concentrations in the crust [13,14]. The latter constitutes the lower ore grade limit. Taking this into account, the concentration exergy of each mineral can be calculated as follows (Equation (1)):

$$b_{c,i} = -RT^0 \left[\ln x_i + \frac{(1 - x_i)}{x_i} \ln(1 - x_i) \right] \tag{1}$$

where x_i is the concentration of substance i, R is the gas constant (8.314 J/molK) and T^0 is the reference temperature (298.15 K). Note, this formula is only valid for ideal gas mixtures and when no chemical cohesion among the substances exists. It is valid for solid mixtures.

The difference obtained between the concentration exergies of a mineral concentration in a mineral deposit (x_m) and that of average concentration in the Earth's crust, *i.e.*, Thanatia (x_c) is the minimum amount of energy that Nature had to spend to concentrate the minerals in a deposit (Equation (2)).

$$\Delta b_c = b_c(x = x_c) - b_c(x = x_m) \tag{2}$$

The exergy replacement costs (b^*) are calculated as follows (Equation (3)):

$$b^* = k(x_c) \cdot \Delta b(x_c \rightarrow x_m) \tag{3}$$

where k is a dimensionless variable that represents the unit exergy cost of a mineral, defined as the ratio between the energy invested in the real obtaining process for mining and concentrating the mineral, and the minimum theoretical energy required if the process from the ore to the final product was reversible. Therefore, k is a measure of the irreversibility of man-made processes and amplifies maximum exergy by a factor of ten to several thousand times, depending on the commodity analyzed.

Another way to understand the relevance of mineral capital losses, caused by domestic extraction, is to convert exergy replacement costs into money, through current energy prices. Together with the DMC ratio, of Equation (4), the impact that economic costs associated with mineral depletion may have on GDP and sustainable development can be assessed.

$$DMC \ ratio = \frac{domestic \ extaction \ (DE)}{extraction \ + \ imports - exports} \tag{4}$$

Note that Equation (4) can be calculated in terms of mass or exergy units, and DE can be replaced with import or export flows to evaluate their respective ratios.

The monetary cost of reversing the extractive process using current energy quantities and prices, can be represented in Equation (5):

$$Monetary \ costs \ of \ ERC \ = \ b^* \cdot p \tag{5}$$

where, p corresponds to the national or world market price of the total amount of energy, from a given source (coal, oil, gas, electricity), required to reverse the mining and beneficiation process and effectively place a mineral back into its original deposit, using current best available technology.

In the case of fossil fuels, economic costs can be directly calculated using their corresponding market prices of the year under consideration. The prices used are the price per barrel in the case of oil, the UK Heren NBP Index price for natural gas and the Northwest Europe marker price for coal [15,16]. For non-fuel minerals the authors considered a range. The lower boundary is calculated assuming that coal is used to re-concentrate mineral capital. The upper boundary calculates with electricity. Electricity prices were obtained from the national statistics services of both respective countries.

The minerals considered are presented in Table 1. The detailed methodology for obtaining the non-fuel mineral data, included in this table, is described in [17]. In the case of fossil fuels, since once they are consumed and burned they cannot be replaced, their exergy content corresponds to their high heating values [18].

Table 1. Exergy replacement costs of the minerals considered in this study [17]).

Substance	Mineral Ore	Exergy Replacement Costs (GJ/ton)	Minerals Analysed for Colombia	Minerals Analysed for Spain
		Non fuels		
Aluminium	Gibbsite	627.3		X
Antimony	Stibnite	474.5		X
Arsenic	Arsenopyrite	399.8		X
Bismuth	Bismuthinite	489.2		X
Cadmium	Greenockite	5898.4		X
Chromium	Chromite	4.5		X
Cobalt	Linaeite	10871.9		X
Copper	Chalcopyrite	110.4	X	X
Fluorspar	Fluorite	182.7		X
Gold	Native gold	583668.4	X	X
Gypsum	Gypsum	15.4		X
Iron ore	Hematite	17.8	X	X
Lead	Galena	36.6		X
Limestone	Calcite	2.6	X	X
Lithium	*Li* in brines	545.8		X
Manganese	Pyrolusite	15.6		X
Mercury	Cinnabar	28,298.0		X
Molybdenum	Molybdenite	907.9		X
Nickel	Pentlandite	761.0	X	X
Phosphate rock	Fluorapatite	0.4		X
Potassium	Sylvite	1224.2		X
Silicon	Quartz	0.7		X
Silver	Argentite	7371.4	X	X
Sodium	Halite	44.1		X
Tin	Cassiterite	426.4		X
Titanium	Ilmenite	4.5		X
Uranium	Uraninite	901.4		X
Wolfram	Scheelite	7429.3		X
Zinc	Sphalerite	24.8		X
Zirconium	Zircon	654.4		X
		Fossil fuels		
Coal		24.3–31.6	X	X
Natural Gas		39.4	X	X
Oil		44.0–46.3	X	X

Data categories include the following: mineral capital loss includes extraction, imports, exports and recycling for 2011, as a common reference year [19,20].

Colombian data is taken predominately from the Colombian Mining Information System (SIMCO) and Ministry of Mines and Energy (Ministerio de Minas y Energía). Copper recycling was taken from the National Register for the Generation of Hazardous Waste (Registro Nacional de Generadores de Residuos Peligrosos). For Spain, the data comes from the Ministry of Industry, Energy and Tourism of Spain (Ministerio de Industria, Energía y Turismo) and Spanish Statistical Office (Instituto Nacional de Estadística) The import/export information was taken from the Chamber of Commerce (Cámara de Comercio). Spanish mineral recycling rates were obtained from a report published by the United Nations Environment Programme [21]. United States Geological Survey (U.S.G.S.) statistics were used as supplementary information for both Colombia and Spain [22,23].

The input and output flows were represented in a Sankey diagram to support visual understanding. Note, even though recycling can be considered both an input and an output, such flows, in this case, were considered only as an output, given that the analysis is carried out for a single year and subsequently it is not logical to assume that the amount recycled in that year was produced that same year.

3. Colombian and Spanish Mineral Resources

3.1. Colombia

In 2011, 40% of Colombian land had either been licensed or solicited for mining concessions [24]. Colombia represented the fifth largest economy in Latin America in 2011 [25]. In addition, 11.3% of the GDP corresponded to the mining sector; mineral export specifically, in monetary value, accounted for 55% of the country's total exported goods in 2011 [26].

According to British Petroleum [16], Colombian proven reserves of oil and natural gas both represent 0.1% respectively of the global reserves. Proven coal reserves, meanwhile, correspond to 0.8%. In 2011, Colombia was the number one coal producer in Latin America, the tenth producer of coal globally and the fourth largest thermal coal exporter [22].

Regarding non-fuel minerals, the country is the number one producer of emeralds in the world, the number one producer of nickel and coal in South America, the only producer of platinum of Latin America and the tenth largest producer of gold in the world. In 2011, non-fuel mineral production was led by limestone (97%) and iron (1.2%) followed by nickel, copper, gold, silver and platinum (accounting for 0.9% in total) [27].

3.2. Spain

In 2011, Spain was the third global producer of gypsum and the sixth producer of fluorspar and Spain's mining and mineral processing industries contributed to 0.8% of Spain's GDP [23].

According to British Petroleum [16], the country does not have proven reserves of oil nor natural gas but does have 530 million tonnes of coal reserves, equivalent to less than 0.1% of world coal reserves. Spain has an almost negligible extraction of natural gas and oil, leaving coal as the most important fuel. In 2011, approximately 6.7 million tonnes of anthracite, bituminous and subbituminous coal were extracted.

The highest level of extraction of any mineral is that of limestone, with 129.6 million tonnes extracted in 2011, a significant amount when compared to the 7.8 million tonnes of gypsum extracted that same year [28]. Regarding metals, there are still some mines, in the northern and eastern zones of the country that extract copper, gold, lead, silver and zinc.

4. Comparative Case Studies: Colombia and Spain

Colombia is investigated due to the national and global developments occurring in its extractive industry and its desire to export. Mining and hydrocarbons form an important part of the drive for development, evidenced by the government's vision to become a mining nation by 2019 [29]. Spain was selected following a general analysis of European Union mineral trade, where it was shown be an average and representative importing and consuming Member State [30].

4.1. Analysis in Unit Tonnes

4.1.1. Colombia

As seen in Figure 1A, Colombian production and export predominate over imports, recycling and consumption. Fossil fuel extraction represents 91.2% of the total. Limestone is the second most extracted mineral, accounting for 8.6%, and contributing, albeit in small quantities, to the export market. The remainder (0.2%) corresponds to iron, nickel, copper, gold, silver and PGM (platinum group metals). Fossil fuel imports represent less than 0.01% of the total mineral balance whilst the contribution of recycling is almost negligible.

Figure 1B is obtained upon removing both fossil fuels and limestone data from Figure 1A. It is subsequently much easier to see that there is an appreciable consumption of iron and that nickel is the most important exported metallic mineral. Even when fossil fuels and limestone are removed from the Sankey diagram, copper, gold, PGM and silver production is still not large enough to register.

4.1.2. Spain

Figure 1C is the graphical representation of the Spanish mineral balance. Limestone predominates over the rest of the materials, representing 86%, in mass terms, of the total fuel and non-fuel mineral production. Imports represent 19.3% of the global mineral balance, and correspond to that of oil, natural gas and aluminium. Recycling and exports do not play a significant role, accounting for only 2.6% between them. If, as before with Colombia, data related to fossil fuels and limestone is removed from Figure 1C, to create Figure 1D, gypsum, salt, potash and feldspar were the most extracted substances in 2011. Gypsum remains one of the most extracted and exported substances in the country, although a little more than half of the production is consumed domestically. Metals, such as aluminium, iron, copper and zinc, are imported.

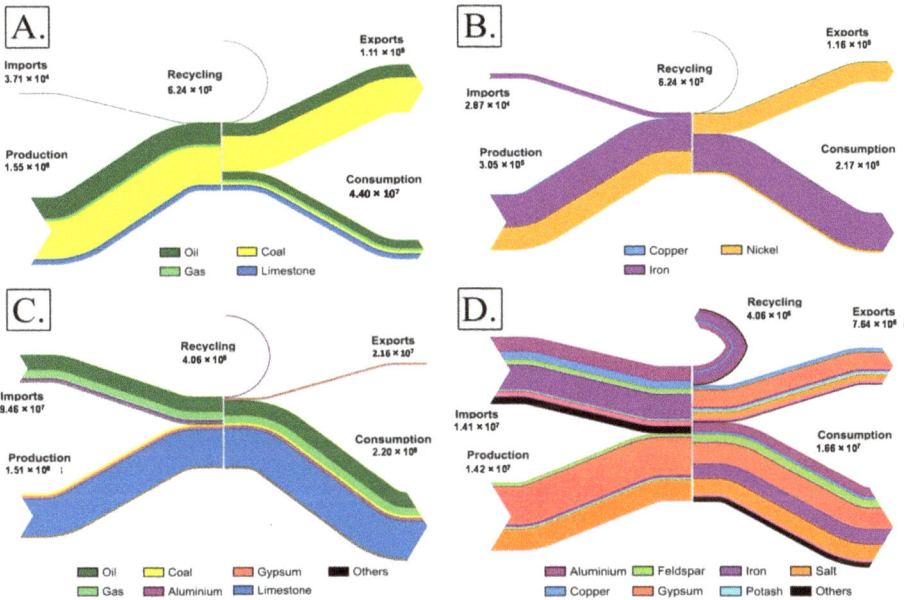

Figure 1. Sankey diagrams representing mineral balance for the year 2011 expressed in tonnes. (**A**) Colombian mineral balance; (**B**) Colombian mineral balance without fossil fuels and limestone; (**C**) Spanish mineral balance; (**D**) Spanish mineral balance without fossil fuels and limestone.

4.2. Comparative Analysis in Exergy Replacement Costs

4.2.1. Colombia

If the 2011 mineral balance is represented, not in mass, but instead in ERC terms (Figure 2A), oil and coal remain the most important Colombian commodities. Oil extraction, measured in percentage mass represents 31.1% of the total. This percentage increases to 43.3%, if oil extraction is measured using exergy replacement costs. The same thing happens for gas extraction (going from 4.6% to 7.2%), and coal extraction changes too (going from 55.4% to 47.6%). Exports, for their part, correspond to 73% of all mineral resources mined nationally.

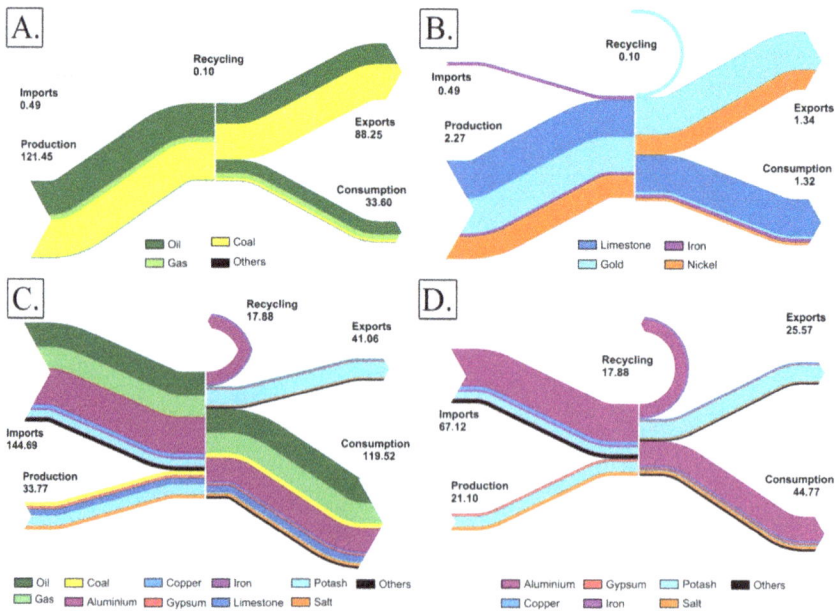

Figure 2. Sankey diagrams representing mineral balance for 2011 expressed in exergy replacement costs. (**A**) Colombian mineral balance; (**B**) Colombian mineral balance without fossil fuels; (**C**) Spanish mineral balance; (**D**) Spanish mineral balance without fossil fuels and limestone.

If fossil fuel data are discarded from the Sankey diagram (to create Figure 2B), the iron or nickel that played an important role in unit mass, become less relevant when considered in exergy replacement terms. Gold, almost exclusively extracted for the export market, occupies a much more significant position. Its mass is almost negligible, but its exergy replacement cost constitutes 35.3% of the total, as it requires more energy intensive processes, and thus is more difficult, to concentrate (or re-concentrate) than say limestone, iron or copper. Comparing the average energy intensity of copper and gold, in the former it is 22.2 GJ per tonne, and, in the case of gold, it is 143 GJ per kilo [31,32]. The importance of limestone extraction thus decreases. It contributed to 97.7% of the non-fuel mineral mass balance but only 37.8% of the exergy replacement costs. This is caused by the predominance of fossil fuel extraction and export, which masks the relevance of other minerals. If fossil fuel data are removed from the Colombian mineral balance, 0.21% of the total comes from imports, when expressed in tonnes, and 17.58 % when presented as exergy replacement costs. Iron is the most imported metal.

4.2.2. Spain

The significance of both oil and gas imports is identical, whether expressed as exergy replacement costs (Figure 2C) or mass (Figure 1C). The most striking variation corresponds to limestone. Limestone extraction measured 86% of the total domestic extraction in mass units but decreases to 23% when using exergy replacement costs. Exports correspond to 23% of the total output minerals. Potash, gypsum and salt (halite) are the most exported commodities.

Aluminium, which has the highest level of import and consumption of any metal in Spain, experiences a notable difference. In terms of percentage in tonnes, it contributed little. Represented through exergy replacement costs, however, and aluminium accounts for 32% of the total import. If, as before, fossil fuels and limestone data are removed from the scenario (Figure 2D), the role of potash and halite in national production and export becomes more apparent.

A summary of the percentage total in mass terms and in exergy replacement costs of production, imports, exports, recycling and consumption for both countries can be found in the Table 2.

Table 2. Percentages for production, imports, exports, recycling and consumption. 2011 data expressed in mass terms and in exergy replacement costs for Colombia and Spain.

Balance Stage	Flow	Colombia (mass)	Colombia (ERC)	Spain (mass)	Spain (ERC)
Inputs	Production	99.98	99.60	61.42	18.92
	Imports	0.02	0.40	35.58	81.08
Outputs	Exports	71.57	72.37	8.80	23.01
	Recycling	<0.01	0.08	1.65	10.02
	Consumption (P+I-E-R)	28.42	27.55	89.54	66.97

In both countries, mineral recycling is extremely low, accounting for less than 0.01% and 1.65% in Colombia and Spain respectively in mass terms. Comparing the mineral balance in mass terms, one can see that the Colombian economy is one of export (71.57%). In 2011, 3.9% of its total exports went to Spain. In the period January to December, Spanish imports of Colombian minerals increased by 204.4%, mainly due to fuels and their derived products [33]. In fact, Spain depends almost entirely on imported fossil fuel supply and almost everything that is imported or Spanish produced is consumed within the country's borders.

On comparing the values in mass and exergy replacement costs terms, it becomes clear that Colombian net values remain within the same magnitude. For Spain, the difference between mass and exergy replacement percentages are noticeable, since Spanish mineral imports consist essentially of those minerals that require highly energy intensive processes to re-concentrate them from a dispersed state (*i.e.*, Thanatia) back into the original natural deposit (*i.e.*, the condition of mineral's existence before having been mined).

4.3. Domestic Mineral Resource Dependency Ratio

For Colombia, the domestic mineral resource dependency ratio (DE/DMC), in the case of non-fuel minerals, is greater than one, indicating that the country is self-sufficient for the minerals analysed, as evidenced by Figures 1A and 2A. In the case of I/DMC, Colombia depends on an external supply of iron, which is subtly identified by the value of exergy replacement costs. For E/DMC, as stated before, gold and nickel were the main minerals exported, and this higher quality, as opposed to quantity, is the main characteristic reflected in these results.

In the case of Colombian fossil fuels, the DE/DMC ratio, when mass is used, is high. This is because, and as aforementioned, Colombia is an exporting country—around 78% of the fossil fuels extracted in 2011 were sent abroad. The DE/DMC ratio decreases by approximately 20% when expressed in exergy replacement costs. As shown in Figure 1A, coal has a significant role in Colombian exports and since all types of coal have lower high heating values than either oil or natural gas, the total exergy value is lower than when expressed in mass terms.

Since the order of magnitude is noticeable, and in an effort to maintain representativeness in the results, the data for non-fuel minerals and fuel minerals has been kept separate, as shown in Table 3.

Table 3. Ratios between domestic extraction and domestic material consumption (DE/DMC), imports to DMC (I/DMC) and exports to DMC (E/DMC) for Colombia and Spain for the year 2011. Data are expressed in mass terms and in exergy replacement costs.

Country	Minerals Analysed	DE/DMC		I/DMC		E/DMC	
		Mass	Exergy	Mass	Exergy	Mass	Exergy
Colombia	Non-fuel minerals	1.40	2.14	0.00	0.02	0.01	1.26
	Fossil fuels	4.63	3.69	0.00	0.00	3.63	2.69
Spain	Non-fuel minerals	0.97	0.45	0.10	1.32	0.05	0.54
	Fossil fuels	0.08	0.07	0.93	1.13	0.16	0.20

As for Spain, and in the case of non-fuel minerals, when domestic mineral resource dependency is considered in mass terms, it appears not to depend highly on import (0.97; with one or higher indicative of a low domestic mineral resource dependency). However, this value reflects the fact that imported non-fuel minerals are masked, in mass terms, by the sizable quantity of construction minerals (in terms of tonnes) that Spanish industry provides for its own market demand. Spanish limestone and gypsum, for example, account for 96.5% of the total.

When domestic mineral resource dependency is calculated using ERC, DE/DMC ratio drops by slightly less than half (0.45). This, along with the elevated value of the I/DMC ratio in exergy terms, indicates that Spain relies on scarcer mineral imports, such as those of aluminium, copper or iron. These minerals have higher exergy replacement costs.

For the export dependency ratio (E/DMC), the most significant 2011 mineral export, in terms of tonnes, was potash. Given its elevated exergy replacement cost, its role becomes more noticeable when considered in exergy rather than mass units.

Since Spanish domestic extraction of fossil fuels is almost negligible, the DE/DMC ratio confirms that Spain depends strongly on external sources, which is also demonstrated by the high I/DMC ratio.

On further analysis, Colombian coal export dependency could present a problem should the market turn to cleaner fuels. Precious metal demand places great pressure on existing reserves and could create future development issues. For Spain, the ERC show a high import dependency on scarcer minerals. The country should either increment domestic production for these minerals or reduce/adapt the activities that require an intensive use of rarer elements.

4.4. Economic Analysis for Colombian and Spanish Mineral Balance

According to Table 4 and Figure 3, if Colombia was charged with having to re-concentrate those minerals that it had extracted in 2011, the GDP of $38 billion (USD, 2011 price) generated by its extractive industries would not cover the ERC monetary equivalent of approximately $56–59 billion (USD), regardless of the energy source used.

Table 4. Monetary costs associated to the mineral extraction in Colombia and Spain for 2011.

Parameter		Colombia	Spain
Exergy (total, Mtoe)		121.45	33.77
Monetary costs (billion $)	*Lower boundary*	55.94	7.63
	Upper boundary	59.40	29.48
% GDP	*Lower boundary*	16.67%	0.52%
	Upper boundary	17.69%	2.02%

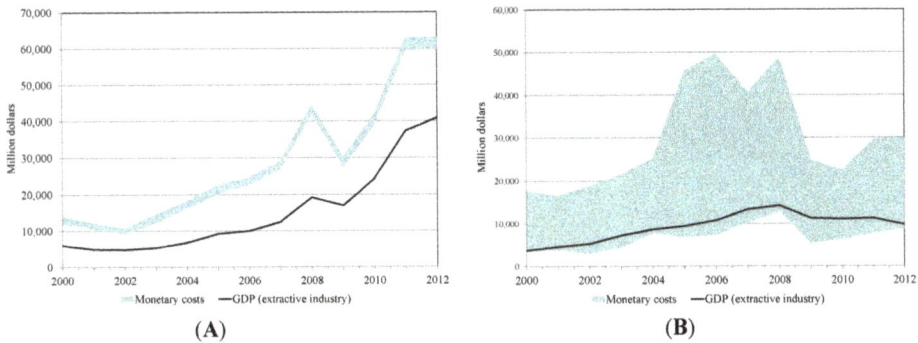

Figure 3. Monetary equivalent of ERC and GDP generated by the extractive sector for Colombia (**A**) and Spain (**B**) from 2000 to 2012.

For Spain, the contribution of the extractive industry was approximately $11.2 billion (USD). This is substantially less than the theoretical amount required to re-concentrate the extracted assets, upon considering the upper boundary. This issue appeared in 2009, as studied by Valero *et al.* [34] who discovered that, on average, prices were 39 times lower than what they needed to be in order to reflect the loss of mineral capital, calculated via exergy replacement costs.

5. Conclusions

A DMC analysis, with exergy replacement costs, makes it possible to evaluate not only the quantity but the quality of international flows transferred between nations. This allows for a more complete picture of mineral dependency and sustainable development.

The economic analysis indicates that Colombia would have lost a monetary equivalent of $56 to $59 billion (USD), if it had to re-concentrate its mineral deposits using energy derived from either coal or electricity, a figure higher than the Colombian mining sector's contribution of $38 billion (USD). Spain's alleged losses would be considerably lower than Colombia's but still between $7 and $29 billion (USD). That said, given the Spanish reliance on imports, most of the country's ERC would be allocated to Colombia and the other trade partners, which supply the mineral demand beyond Spain's own domestic production capabilities.

If exergy replacement costs are considered, the theoretical figures presented in this paper can link market prices to values that more readily correlate to physical costs. Currently, the fact that the true cost of mineral resource extraction is not considered in ERC terms means that current rates of extraction at 2011 prices are not sustainable.

Acknowledgments: This study has been carried out under the framework of the ENE2014-59933-R project, financed by the Spanish Ministry of Economy and Competitiveness.

Author Contributions: These authors contributed equally to this work.

Conflicts of Interest: The authors declare no conflict of interest.

References

1. Krausmann, F.; Gingrich, S.; Eisenmenger, N.; Erb, K.H.; Haberl, H.; Fischer-Kowalski, M. Growth in global materials use, GDP and population during the 20th century. *Ecol. Econ.* **2009**, *68*, 2696–2705. [CrossRef]
2. Eurostat. Glossary: Domestic Material Consumption (DMC). Available online: http://ec.europa.eu/eurostat/statistics-explained/index.php/Glossary:Domestic_material_consumption_(DMC) (accessed on 30 October 2015).

3. Weisz, H.; Krausmann, F.; Amann, C.; Eisenmenger, N.; Erb, K.H.; Fischer-Kowalski, M.; Hubacek, K. The physical economy of the European Union: Cross-country comparison and determinants of material consumption. *Ecol. Econ.* **2006**, *58*, 676–698. [CrossRef]
4. Valero, A.; Valero, A.; Calvo, G. Using thermodynamics to improve the resource efficiency indicator GDP/DMC. *Resour. Conserv. Recycl.* **2015**, *94*, 110–117. [CrossRef]
5. British Geological Survey. Risk list 2012. Available online: https://www.bgs.ac.uk/downloads/start.cfm?id=2643 (accessed on 11 November 2015).
6. European Commission. Report on Critical Raw Materials for the EU. Report of the Ad hoc Working Group on Defining Critical Raw Materials. Available online: https://ec.europa.eu/growth/tools-databases/eip-raw-materials/en/community/document/critical-raw-materials-eu-report-ad-hoc-working-group-defining-critical-raw (accessed on 30 October 2015).
7. Henckens, M.L.C.M.; Driessen, P.P.J.; Worrell, E. Metal scarcity and sustainability, analyzing the necessity to reduce the extraction of scare metals. *Resour. Conserv. Recycl.* **2014**, *93*, 1–8. [CrossRef]
8. Valero, A. Thermoeconomics as a conceptual basis for energy-ecological analysis. In *Advances in Energy Studies*; Ulgiati, S., Ed.; Musis: Roma, Italy, 1998; pp. 415–444.
9. Ayres, R.U. Eco-thermodynamics: Economics and the second law. *Ecol. Econ.* **1998**, *26*, 189–209. [CrossRef]
10. Sciubba, E. "Exergo-economics: Thermodynamic foundation for a more rational resource use". *Int. J. Energy Res.* **2005**, *29*, 613–636. [CrossRef]
11. Szargut, J.; Morris, D.R.; Steward, F.R. *Exergy Analysis of Thermal, Chemical, and Metallurgical Processes*; Hemisphere Pub. Co.: New York, NY, USA, 1988.
12. Valero, A.; Valero, A. From grave to cradle. A thermodynamic approach for accounting for abiotic resource depletion. *J. Ind. Ecol.* **2013**, *17*, 43–52. [CrossRef]
13. Valero, A.; Valero, A.; Gómez, J.B. The crepuscular planet. A model for the exhausted continental crust. *Energy* **2011**, *36*, 694–707. [CrossRef]
14. Valero, A.; Agudelo, A.; Valero, A. The crepuscular planet. A model for exhausted atmosphere and hydrosphere. *Energy* **2011**, *36*, 3745–3753. [CrossRef]
15. BP Global. *British Petroleum Statistical Review of World Energy*; BP Global: London, UK, June 2012; p. 48.
16. BP Global. *British Petroleum Statistical Review of World Energy*; BP Global: London, UK, June 2014; p. 48.
17. Valero, A.; Valero, A. *Thanatia: The Destiny of the Earth's Mineral Resources: A Thermodynamic Cradle-to-Craddle Assessment*; World Scientific Publishing Company: Singapore, 2014.
18. Valero, A.; Valero, A. What are the clean reserves of fossil fuels? *Resour. Conserv. Recycl.* **2012**, *68*, 126–131. [CrossRef]
19. Carmona, L.G.; Whiting, K.; Valero, A.; Valero, A. Colombian Mineral Resources: An Analysis from a Thermodynamic Second Law Perspective. *Resour. Policy* **2015**, *45*, 23–28. [CrossRef]
20. Calvo, G.; Valero, A.; Valero, A.; Carpintero, Ó. An exergoecological analysis of the mineral economy in Spain. *Energy* **2015**, *88*, 2–8. [CrossRef]
21. Graedel, T.E.; Allwood, J.; Birat, J.-P.; Reck, B.K.; Sibley, S.F.; Sonnemann, G.; Buchert, M.; Hagelüken, C. (Eds.) *Recycling Rates of Metals—A Status Report*. Available online: http://www.unep.org/resourcepanel/Portals/24102/PDFs/Metals_Recycling_Rates_110412-1.pdf (accessed on 11 November 2015).
22. Wacaster, S. United States Geological Survey: 2012 Minerals Yearbook. Available online: http://minerals.usgs.gov/minerals/pubs/country/2012/myb3-2012-co.pdf (accessed on 11 November 2015).
23. Gurmendi, A.C. United States Geological Survey: The mineral industry of Spain (2011). 2013; p. 11. Available online: http://minerals.usgs.gov/minerals/pubs/country/2011/myb3-2011-sp.pdf (accessed on 11 November 2015).
24. Sanchez-Garzoli, G. *Stopping irreparable harm: Acting on Colombia's Afro-Colombian and Indigenous Communities Protection Crisis*. Available online: http://www.peacebuilding.no/var/ezflow_site/storage/original/application/04fcd8f818b16e1c31c4306ad74dfb70.pdf (accessed on 11 November 2015).
25. Rebossio, A. ¿Cuál es la tercera economía latinoamericana? ¿Y la sexta? El Pais Colombia 23/12/2011. Available online: http://blogs.elpais.com/eco-americano/2011/12/cu%C3%A1l-es-la-tercera-econom%C3%ADa-latinoamericana-y-la-sexta.html (accessed on 11 November 2015).
26. Wacaster, S. United States Geological Survey: 2011 Minerals yearbook, Colombia. Available online: http://minerals.usgs.gov/minerals/pubs/country/2011/myb3-2011-co.pdf (accessed on 11 November 2015).

27. Sistema de Información Minero Colombiano (SIMCO). Producción Oficial de Minerales en Colombia. Available online: http://www.simco.gov.co/simco/Estad%C3%ADsticas/Producci%C3%B3n/tabid/121/Default.aspx?PageContentID=99 (accessed on 30 October 2015).

28. Ministerio de Industria, Energía y Turismo. Estadística minera: 2011. Available online: http://www.minetur.gob.es/energia/mineria/Estadistica/2011/anual%202011.pdf (accessed on 11 November 2015).

29. UPME. Colombia País Minero: Plan Nacional para el Desarrollo Minero, Visión 2019. Subdirección de Planeación Minero. Available online: http://www.upme.gov.co/Docs/PNDM_2019_Final.pdf (accessed on 30 October 2015).

30. Calvo, G.; Valero, A.; Valero, A. Material flow analysis for Europe: An exergoecological approach. *Ecol. Indic.* **2016**, *60*, 603–610. [CrossRef]

31. Mudd, G. Global trends in gold mining towards quantifying environmental and resource sustainability. *Resour. Policy* **2007**, *32*, 42–56. [CrossRef]

32. Northey, S.; Haque, N.; Mudd, G. Using sustainability reporting to assess the environmental footprint of copper mining. *J. Clean. Product.* **2013**, *40*, 118–128. [CrossRef]

33. Departamento Administrativo Nacional de Estadística (DANE). Boletín de prensa: Comercio exterior, exportaciones y balanza comercial de diciembre de 2011. Available online: https://www.dane.gov.co/files/investigaciones/boletines/exportaciones/bol_exp_dic11.pdf (accessed on 11 November 2015).

34. Valero, A.; Carpintero, Ó.; Valero, A.; Calvo, G. How to account for mineral depletion. The exergy and economic mineral balance of Spain as a case study. *Ecol. Indic.* **2014**, *46*, 548–559. [CrossRef]

Article

Surplus Cost Potential as a Life Cycle Impact Indicator for Metal Extraction

Marisa D.M. Vieira [1,2,*], Thomas C. Ponsioen [2], Mark J. Goedkoop [2] and Mark A.J. Huijbregts [1]

[1] Radboud University Nijmegen, Faculty of Science, Department of Environmental Science, P.O. Box 9010, Nijmegen 6500 GL, The Netherlands; m.huijbregts@science.ru.nl

[2] PRé Consultants b.v., Stationsplein 121, Amersfoort 3818 LE, The Netherlands; ponsioen@pre-sustainability.com (T.C.P.); goedkoop@pre-sustainability.com (M.J.G.)

* Correspondence: vieira@pre-sustainability.com

Academic Editor: Mario Schmidt
Received: 28 October 2015; Accepted: 22 December 2015; Published: 6 January 2016

Abstract: In the evaluation of product life cycles, methods to assess the increase in scarcity of resources are still under development. Indicators that can express the importance of an increase in scarcity of metals extracted include surplus ore produced, surplus energy required, and surplus costs in the mining and the milling stage. Particularly the quantification of surplus costs per unit of metal extracted as an indicator is still in an early stage of development. Here, we developed a method that quantifies the surplus cost potential of mining and milling activities per unit of metal extracted, fully accounting for mine-specific differences in costs. The surplus cost potential indicator is calculated as the average cost increase resulting from all future metal extractions, as quantified via cumulative cost-tonnage relationships. We tested the calculation procedure with 12 metals and platinum-group metals as a separate group. We found that the surplus costs range six orders of magnitude between the metals included, *i.e.*, between $0.01–$0.02 (iron) and $13,533–$17,098 (rhodium) USD (year 2013) per kilogram of metal extracted. The choice of the reserve estimate (reserves *vs.* ultimate recoverable resource) influenced the surplus costs only to a limited extent, *i.e.*, between a factor of 0.7 and 3.2 for the metals included. Our results provide a good basis to regularly include surplus cost estimates as resource scarcity indicator in life cycle assessment.

Keywords: characterization factors; endpoint; life cycle assessment; metals; mining; resource scarcity

---

1. Introduction

The need for reduction of greenhouse gas emissions will most likely shift energy generation from fossil fuels to alternative sources of energy such as solar, wind and nuclear power [1]. Renewable energy production technologies often require significantly more metals in their construction, such as copper, indium, lead, and molybdenum, compared to fossil energy [2]. It is, therefore, important to understand the trade-offs in using different types of mineral, metal, and fossil resources. Life cycle assessment (LCA) is a method that is capable of quantifying trade-offs in terms of potential impacts on human health, natural environment and natural resources [3]. Particularly, life cycle impact assessment (LCIA) methods assessing impacts on natural resources are still in an early stage of development [4–6].

Many indicators previously developed for resource use in LCA focused on evaluating the exhaustion of resources in the earth. Tilton [7] defined this as the fixed stock paradigm, which implies that resources are limited and will eventually deplete. However, physical availability is most likely not the main problem as many metals are not destroyed after their use, but their economic availability is [8–12]. Even though there may be commodities available to mine in the Earth's crust, the cost of extraction can be higher than the price consumers are willing to pay [7,9–14]. For fossil resources the same is observed, with costs being the limiting factor for fossil resource extraction [15]. Since costs appear to be the limiting factor for the use of both metal and fossil resources, a resource scarcity indicator expressed in costs for these resources is considered worthwhile to investigate [11,12,16–19].

The European Commission-Joint Research Centre—Institute for Environment and Sustainability [5] indicated surplus costs, as explored by the ReCiPe method [20], as a promising indicator to assess resource use among the existing approaches. However, its application is not yet considered mature for recommendation [5]. For instance, mining and milling costs per unit of ore extracted were considered constant across all mines [20]. Surplus costs are defined as the (economic) burden that current resource extraction puts on future situations. These surplus costs over a long period of time are not particularly considered in current decision making [10]. This definition closely links up with the focus on externalities in LCA, *i.e.*, the cost that affects a party who did not choose to incur that cost [21,22]. The quantification of these external costs directly fit into the framework of environmental life cycle costing (LCC) [23,24].

The goal of this paper was to develop a method that quantifies the surplus cost potential per unit of metal extracted in the future, fully accounting for co-production and mine-specific differences in costs. We demonstrated how the method can be applied in practice for 12 metal commodities including uranium, which is also an energy resource, and platinum-group metals as a group.

2. Methods and Data

2.1. Cause-Effect Pathway

Metals may come from two origins, either from mining or from secondary production, the latter resulting from waste recovery. The current fraction of secondary production has to be included in the life cycle inventory and is therefore not considered in the LCIA step. Without technological development in the mining industry, increased primary metal extraction results in a subsequent increase in mining and milling costs, as mines with lower operating costs are explored first. The operating costs account for co-production and are allocated across all mine products in proportion to their revenue to the mine operator. The average cost increase resulting from all future extraction of a metal is calculated to arrive at the surplus costs per unit of metal extracted.

2.2. Cumulative Cost-Tonnage Relationships

Following the principle that mining sites with lower costs are the first to be explored, mines producing a certain metal were sorted by increasing order of costs per metal extracted. Vieira *et al.* [25] demonstrated that ore grades tend to decrease with the increase of copper extraction and that a log-logistic regression can be applied to cumulative grade-tonnage relationships. Following the same line of reasoning for costs, a log-logistic distribution can be fitted on the inverse of operating costs per metal extracted and the cumulative metal extracted to account for the skewness in the data points [25] by:

$$\frac{1}{C_x} = \exp\left(\alpha_x\right) \cdot \exp\left(\beta_x \cdot \ln\left(\frac{A_{x,\text{sample}} - CME_{x,\text{sample}}}{CME_{x,\text{sample}}}\right)\right) \tag{1}$$

where C_x is the operating cost per metal x produced (in USD/kg$_x$), α_x is the scale parameter, and β_x is the shape parameter of the log-logistic distribution, and $A_{x,\text{sample}}$ and $CME_{x,\text{sample}}$ (both in kg$_x$) are, respectively, the total metal x extracted and the cumulative metal x extracted for the data sample used to derive the curve.

2.3. Characterization Factors

The surplus cost potential per unit of extraction of a metal is an indicator of the future economic scarcity of the extraction of this resource. The characterization factor (CF), defined as the average Surplus Cost Potential (SCP) of metal x, was calculated by:

$$CF_x = \frac{\int_{CME_{x,total}}^{MME_x} C_x\,(ME_x)\,dME_x}{R_x} = \frac{\int_{CME_{x,total}}^{MME_x} C_x\,(ME_x)\,dME_x}{MME_x - CME_{x,total}} \qquad (2)$$

where C_x is the operating cost determined via the cumulative cost-tonnage curve of metal x for an amount of metal x extracted (ME_x), R_x (kg_x) is the reserve of metal x, defined as the difference between all current known cumulative tonnage of metal x extracted ($CME_{x,total}$) and the global maximum tonnage of metal x ever extracted (MME_x). This is illustrated in Figure 1.

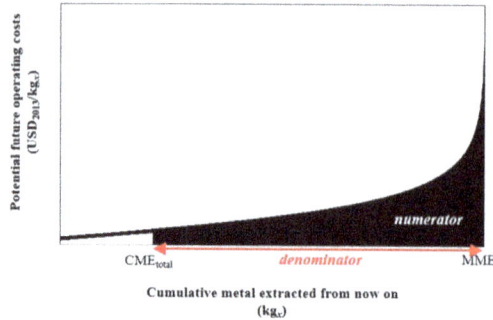

Figure 1. Visualization of the derivation of the surplus cost potential following an average approach and a log-logistic cumulative cost-tonnage curve from the current cumulative metal extracted (CME_{total}) up to global maximum metal extracted (MME).

The overall surplus cost potential caused by a product, e.g., production of photovoltaic electricity in Spain, can be calculated by:

$$IS_{nr} = \sum_x CF_x \times M_x \qquad (3)$$

where IS_{nr} is the impact score of natural resource extraction per functional unit of the product analyzed, e.g., production of 1 kWh of photovoltaic electricity in Spain (USD/kWh solar electricity); CF_x is the characterization factor for resource x extracted (USD/kg_x); and M_x is the amount of resource x extracted per functional unit (e.g., kg_x/kWh solar electricity). This impact score is measured at the endpoint level, thus measuring the potential damage caused, and aggregates the impact score of various types of abiotic natural resources, namely metal, mineral, and fossil resources.

2.4. Sensitivity Analysis

Alternative scenarios were calculated and analyzed with regards to the future production of metals. Since it is unclear what the exact future production of each metal will be, reserves data were used. Reserves data are variable because, on the one hand, ores are constantly being explored and the extraction feasibility may decrease and, on the other hand, new deposits are being discovered or become economically viable. To study the uncertainty of the future production of metals two different reserve estimates were applied. The first type of reserve estimate is the "Reserves (R_R)" which is defined as that part of a mineral resource "which could be economically extracted or produced at the time of determination", meaning at current prices, state of technology, *etc.* [26]. The other type of reserve estimate, the "ultimate recoverable resource (R_{URR})", refers to "the amount available in the upper earth's crust that is ultimately recoverable". The definition of ultimate recoverable resource as used by UNEP [27], there called extractable global resource (EGR), was used here. This is assuming that 0.01% of the total amount in the crust to 3 km depth will ultimately be available.

2.5. Data Selection and Fitting

Characterization factors were derived for copper (Cu), iron (Fe), lead (Pb), manganese (Mn), molybdenum (Mo), nickel (Ni), palladium (Pd), platinum (Pt), rhodium (Rh), silver (Ag), uranium (U), zinc (Zn), and for platinum-group metals (PGM) as a group. Table 1 shows the data for the metals included in the calculation of the characterization factors.

Table 1. Global cumulative extraction and reserve estimates used for deriving the characterization factors expressed as surplus cost potential [26–32].

Metal	Cumulative Metal Extracted CME_{total} (kg_x)	Maximum Metal Extracted MME_R (kg_x)	Maximum Metal Extracted MME_{URR} (kg_x)
Copper	5.92×10^{11}	1.28×10^{12}	5.22×10^{12}
Iron	3.41×10^{13}	1.15×10^{14}	6.49×10^{15}
Lead	2.35×10^{11}	3.24×10^{11}	3.04×10^{12}
Manganese	5.80×10^{11}	1.15×10^{12}	1.28×10^{14}
Molybdenum	6.62×10^{9}	1.76×10^{10}	1.89×10^{11}
Nickel	5.53×10^{10}	1.29×10^{11}	7.82×10^{12}
Palladium	5.15×10^{6}	2.83×10^{7}	8.25×10^{7}
Platinum	6.97×10^{6}	3.82×10^{7}	1.12×10^{8}
Rhodium	7.36×10^{5}	4.04×10^{6}	1.18×10^{7}
Silver	1.13×10^{9}	1.65×10^{9}	2.11×10^{10}
Uranium	2.71×10^{9}	5.23×10^{9}	4.33×10^{11}
Zinc	4.58×10^{11}	7.08×10^{11}	1.16×10^{13}
Platinum-Group Metals	1.47×10^{7}	8.07×10^{7}	2.36×10^{8}

2.5.1. Cumulative Cost-Tonnage Relationships

To derive log-logistic cumulative cost-tonnage relationships, yearly metal production (in kg_x) and mining and milling costs per metal produced data (in USD/kg_x) between 2000 and 2013 were purchased from the World Mine Cost Data Exchange (WMCDE) [33]. The database contained for each metal one cost curve per year with production and cost data for each mine. The data used for deriving the cumulative cost-tonnage relationships are presented in Table S1 in the Supplementary Material.

The log-logistic distribution for each metal was derived as follows:

- Typically, deposits contain various metals but there is often a main metal that justifies the operation of a mine exploring that deposit. As such, the operating costs of a mine are to be shared by all outputs with a market value. In the World Mine Cost Data Exchange [33], the costs were allocated across all mine products in proportion to their production (monetary) value to the mine operator.
- Each table includes data in U.S. dollars valued in the year it represents, e.g., the costs for 2004 are expressed in constant USD valued in 2004. The CPI Inflation Calculator [34] was used to convert all costs into U.S. dollars for 2013 (USD$_{2013}$).
- For each mine, the weighted average costs per amount of metal produced, calculated on basis of the operating costs per metal in that mine each year and the production tonnages for the same years, and the total metal extracted in the period covered for that mine were calculated.
- The mines were then ranked in increasing order of costs per amount of metal extracted and the cumulative metal extracted for each mine was calculated by adding the metal extracted of that mine to that of all previous mines with lower operating costs.
- A log-logistic fit was applied to the inverted costs for every mine with the software R for statistical computing [35] to derive the scale parameter α and the shape parameter β, including their 95% confidence interval (95% CI). The R square (R^2) of the log-logistic fit was also determined. For the derivation of α and β, the total tonnage of metal extracted A_x was set equal to the total metal extracted as reported in the WMCDE database between 2000 and 2012 or 2013, depending on the metal [33].

2.5.2. Characterization Factors

To derive the surplus cost potential per metal x, the cumulative metal extracted up to now ($CME_{x,total}$) is calculated as the total production from Kelly and Matos [28] which contains world mine production tonnage since 1900 until 2012 for every metal under study except for uranium. For uranium, data was retrieved from Nuclear Energy Agency and International Atomic Energy Agency [29].

To calculate the maximum amount of metal x extracted (MME_x), global reserve estimates for each metal are required. As mentioned above, two types of reserves were used: reserves and ultimate recoverable resource. Reserves were set equal to the global mineral reserves as specified by the U.S. Geological Survey [26]. U.S. Geological Survey [26] estimated the reserves for all metals under study, except for the individual platinum-group metals (PGM) and uranium. For uranium, the global reserves, following the same definition, were taken from Hall and Coleman [30]. Reserves fraction of platinum (47.3%), palladium (35.0%), and rhodium (5.0%) were calculated from data for the three PGM ore types mined in the Bushveld deposit situated in South Africa which accounts for nearly 90 percent of PGM's reserves [31].

Schneider *et al.* [32] estimated the ultimate recoverable resource (URR) for 19 metals and for PGM as a group, eight of which are relevant for this study. The URR data for silver and uranium, following the same definition, was taken from UNEP [27]. The data used for the two types of reserves are presented in Table S2 in the Supplementary Material.

3. Results

3.1. Cumulative Cost-Tonnage Relationships

Figure 2 shows the log-logistic relationships between cumulative extraction of metal x and increase in operating costs per unit of metal x. A number of metals (copper, lead, manganese, molybdenum, nickel, palladium, platinum) and PGM indicate relatively high costs with low production for a few mines. This can be seen from Figure 2, as a few data points on the right hand side of the curve are almost in a vertical line, meaning low production tonnage and at the same time substantially higher costs compared to other mines. These costs often result from co-production in which the main metal being extracted is another than the one under study. Taking Figure 2a for copper as an example, the mine with the highest cost per kilogram of copper extracted is Copper Rand in Canada, with gold and silver as co-products [33]. Since 2009, this mine has been closed due to high costs in comparison with the low copper prices [36]. However, if prices rise this mine may be reopened. This effect is in line with the method here proposed in which mines with lower operating costs per output of metal are extracted first.

The log-logistic fit between costs and cumulative extraction of a metal resulted in explained variances between 77% (for palladium) and 97% (for iron, lead, silver, and zinc). The parameters derived for the log-logistic cost-tonnage relationships are presented in Table S3 in the Supplementary Material.

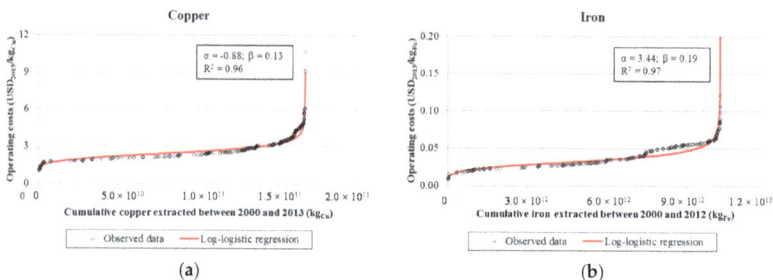

Figure 2. *Cont.*

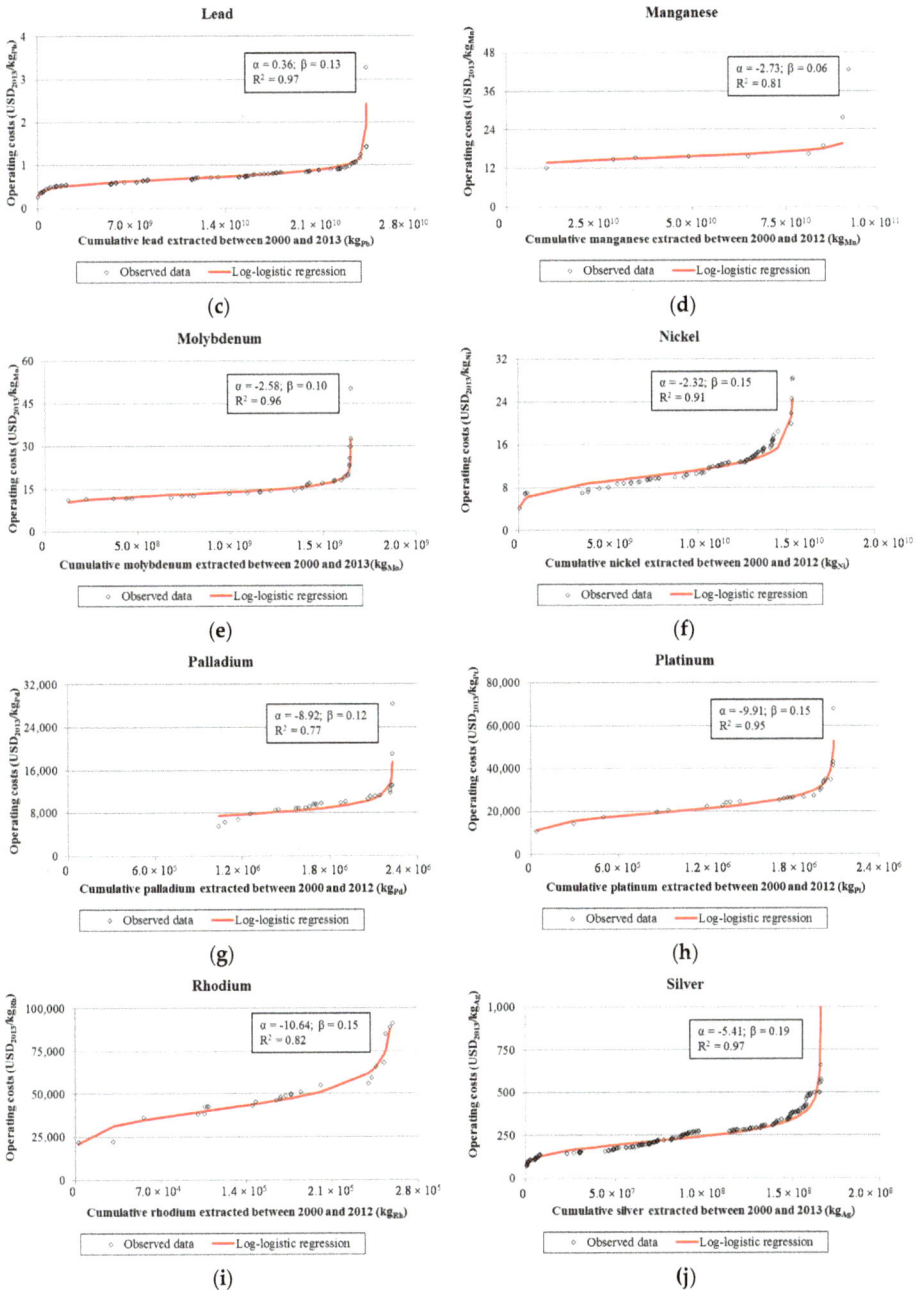

Figure 2. *Cont.*

Uranium

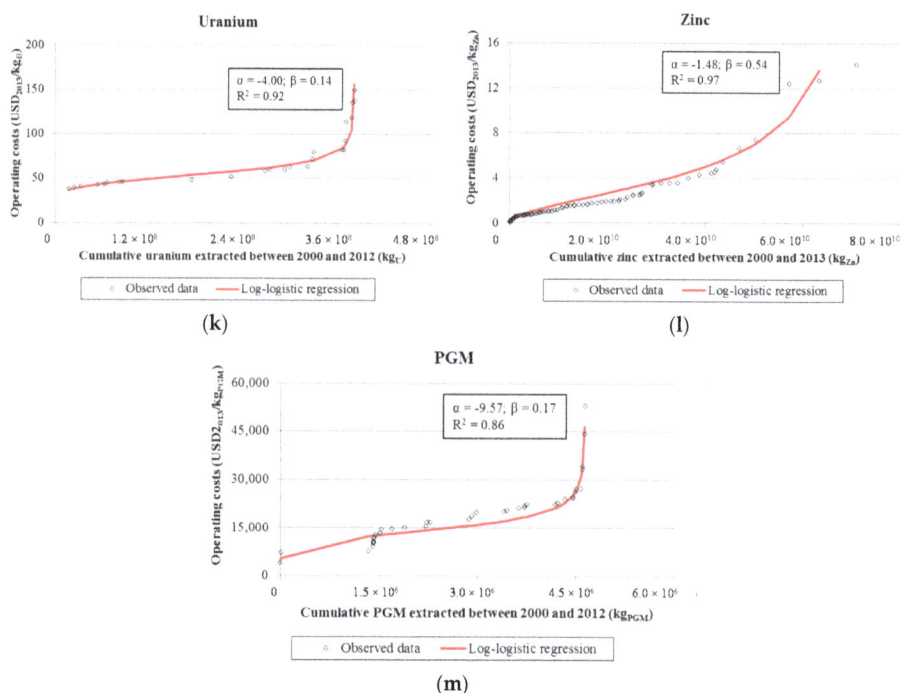

$\alpha = -4.00;\ \beta = 0.14$
$R^2 = 0.92$

Zinc

$\alpha = -1.48;\ \beta = 0.54$
$R^2 = 0.97$

(k)

(l)

PGM

$\alpha = -9.57;\ \beta = 0.17$
$R^2 = 0.86$

(m)

Figure 2. Cumulative cost-tonnage relationships and scale α and shape β derived with the log-logistic distribution for copper (**a**), iron (**b**), lead (**c**), manganese (**d**), molybdenum (**e**), nickel (**f**), palladium (**g**), platinum (**h**), rhodium (**i**), silver (**j**), uranium (**k**), zinc (**l**), and platinum-group metals (**m**). The data were taken from the World Mine Cost Data Exchange [33] for the period from 2000 to 2012/2013.

3.2. Characterization Factors

The characterization factors derived for the metals under study are presented in Figure 3 and can also be found in Table S4 in the Supplementary Material. The CFs obtained vary up to six orders of magnitude. Rhodium resulted in the highest CF of $13,533–$17,098 USD$_{2013}$ per kg of rhodium extracted, while the lowest CF derived was for iron with $0.01–$0.02 USD$_{2013}$ per kg of iron extracted.

With the exception of zinc, the surplus cost potentials derived are always largest using ultimate recoverable resource compared to using the reserves. The CF depends on the combination of the scale parameter (α), the shape parameter (β), CME, and MME of a metal. It is not necessarily true that that the CF gets always higher when MME becomes larger. In fact with very small MME values, so very low reserves, the starting point of the cumulative cost calculation (CME$_{total}$) is in the steepest part of the curve and the CFs become very large. This means that getting the last amount of metal out of the ground will be very costly. Applying ultimate recoverable resource instead of reserves always result in a lower fraction mined of the total amount available another working point. For instance, for copper 46% of the total amount is already mined using reserves, while 11% is mined according to ultimate recoverable resources. For zinc this is 56% *versus* 4%. Depending on the shape (β) of the log-logistic curve (relatively steep for copper and relative shallow for zinc), the increase in costs over the full amount of metal mined can be larger or lower. In this case, the average steepness of the cost curve for copper is larger for 11% to 100% metal extracted compared to 46% to 100%. In contrast, the average steepness of the cost curve for zinc is larger for 56% to 100% metal extracted compared to 4% to 100%.

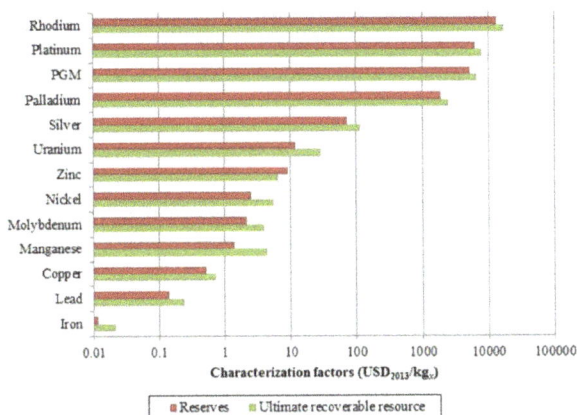

Figure 3. Characterization factors expressed as surplus cost potential (USD$_{2013}$/kg$_x$ extracted) using reserves and ultimate recoverable resource as future extraction estimates.

4. Discussion

4.1. Comparison with other Indicators

Although log-logistic functions can be applied to derive both cumulative cost and ore grade curves, the exact causal relationship between ore grade decrease and surplus costs is unknown as far as we know of. Though ore grade can be considered as an important factor for determining the operating costs in a mine [37], other factors that change over time can be important as well. These are further discussed in the Limitations section.

For all metals under study the current operating costs were derived, calculated as the weighted average operating costs of all mines for 2013 on basis of their yearly production of that metal, from the World Mine Cost Data Exchange [33]. Goedkoop and De Schryver [20] derived characterization factors in the ReCiPe method expressed as surplus costs following a simplified approach of the method here proposed, e.g., with mining and milling costs per unit of ore considered constant across all mines. Steen [38] derived characterization factors in the method to environmental priority strategies for product development (EPS), defined as the costs to be paid to extract the mineral resources using future technologies. For this method, a market scenario was created, where all future generations are included and are imagined to bid on the present abiotic stock reserves [38]. All three monetary results per metal can be seen in Table S5 in the Supplementary Material.

In Figure 4, the current operating costs, the CFs derived in the ReCiPe method with 3% discounting, and the CFs derived in the EPS method are compared to the CFs here derived using reserves as reserve estimate. For all metals except for zinc, the current operating costs are larger than the surplus costs derived here. There is a factor difference of between 0.7 for zinc and a factor of about 16 for manganese. From the log-linear regression in Figure 4, it can be seen that the surplus cost potentials will be about 5.6 (equal to the intercept of the regression) times lower than its current operating costs.

According to the log-linear regression derived in Figure 4 the CFs here derived give typically a factor of 1.3 lower CFs compared to the ReCiPe method with a coefficient of determination (R^2) of 0.81. Note that for individual metals, the differences between the two methods can be much larger, *i.e.*, a factor of seven lower for molybdenum and a factor of 57 higher for zinc in our study compared to the Goedkoop and De Schryver [20] method.

Compared to the EPS method, the CFs derived here give on average a factor of 126 lower with a coefficient of determination (R^2) of 0.88 (see log-linear regression in Figure 4). The differences between the CFs derived by each method for individual metals can differ to up to four orders of magnitude, *i.e.*, a factor of four for manganese and of 4×10^3 for palladium.

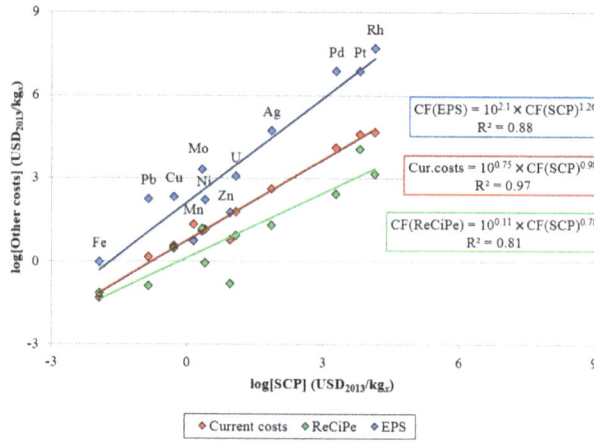

Figure 4. Log-linear relationship between the characterization factors (SCP) here derived, current operating costs [33], the CFs derived in the ReCiPe method [20], and the CFs derived in the EPS method [38]. Both axis are presented in logarithmic scale.

4.2. Limitations

Although we improved the surplus cost calculations of metals compared to current practice, there are still limitations in our calculations as we assumed that costs increase with the increase of metal extracted without explicitly considering the mine type, ore type, new discoveries and technological developments. Currently, open pit mining is preferred due to lower operational costs even though underground mines often contain higher metal grades [39–41]. As open pit mines will most likely become depleted in the future, a shift towards underground mining is likely to occur. Gerst [42] estimated that mining operations have to become deeper than the current open pit mining operations. Furthermore, different ores have different characteristics and moving from one to the other may result in substantially different mining conditions and, as a result, in different operating costs [43]. The cost and tonnage data used to derive the cumulative cost-tonnage curves, however, does not contain information on the mine type and ore type. This is why we did not separately derive cumulative cost-tonnage curves for open pit and underground mines and different ore types in our study. Additionally, there is evidence of overall reduction in mining costs per amount of ore extracted due to technological advances in the mining sector [39,44]. For instance, mines have been increasing in size and the larger the mine size the lower the costs per output extracted due to economies of scale [39,45]. Since we used actual cost data, difference in mine sizes and other differences in technological developments are included in the cumulative cost-tonnage relationships derived. However, the time frame covered by the data is relatively short (13 years). Mining cost data over a longer period of time for deriving the cumulative cost-tonnage relationships would therefore be desirable.

Finally, the method here proposed also assumes that mines with lower operating costs per mine output are mined first. However, there may be both mines with high and low costs operating at the same time and this may be justified by higher metal prices.

5. Conclusions

In this paper, we demonstrated how characterization factors for resource scarcity can be derived, reflecting the surplus cost potential due to a unit increase in metal extraction. Although the surplus cost indicator we propose is measured in monetary units, the surplus cost potential is about quantifying externalities and not about current economic costs as such. The surplus costs are an (economic) burden that we put on future generations resulting from a decrease in highly concentrated and easily accessible resources and can be used in environmental LCC. Our results emphasize the importance of (1) metal-specific cumulative cost-tonnage relationships and (2) future metal extraction estimates. Additional cost and mining data is required to expand the method to include other mineral commodities in evaluating scarcity of mineral commodities in an LCA context.

Acknowledgments: The research was funded by the European Commission under the 7th framework program on environment; ENV.2009.3.3.2.1: LC-IMPACT—Improved Life Cycle Impact Assessment methods (LCIA) for better sustainability assessment of technologies, grant agreement number 243827. M. Vieira was further supported by the Dutch National Institute for Public Health and the Environment RIVM-project S/607020, Measurably Sustainable within the spearhead Healthy and Sustainable Living Environment, commissioned by the Director-General of RIVM, and run under the auspices of RIVM's Science Advisory Board.

Author Contributions: Marisa Vieira is the main author and is responsible for gathering the data and performing the analysis described in this paper. Thomas Ponsioen assisted with the regression analysis. Mark Goedkoop helped to define the research questions. Mark Huijbregts guided the worked, contributed to the interpretation of the results and suggested revisions to original text.

Conflicts of Interest: The authors declare no conflict of interest.

References

1. Kleijn, R. Materials and Energy: A Story of Linkages. Ph.D. Thesis, Leiden University, Leiden, The Netherlands, 5 September 2012.
2. Kleijn, R.; van der Voet, E.; Kramer, G.J.; van Oers, L.; van der Giesen, C. Metal requirements of low-carbon power generation. *Energy* **2011**, *36*, 5640–5648. [CrossRef]
3. Guinée, J.B.; Gorrée, M.; Heijungs, R.; Huppes, G.; Kleijn, R.; de Koning, A.; van Oers, L.; Wegener Sleeswijk, A.; Suh, S.; Udo de Haes, H.A.; *et al. Handbook on Life Cycle Assessment. Operational Guide to the ISO Standards. I: LCA in Perspective. IIa: Guide. IIb: Operational annex. III: Scientific Background*; Kluwer Academic Publishers: Dordrecht, The Netherlands, 2002.
4. Steen, B. Abiotic resource depletion: Different perceptions of the problem with mineral deposits. *Int. J. Life Cycle Assess.* **2006**, *11*, 49–54. [CrossRef]
5. European Commission-Joint Research Centre-Institute for Environment and Sustainability. *International Reference Life Cycle Data System (ILCD) Handbook—Recommendations for Life Cycle Impact Assessment in the European Context*, 1st ed.; Publications Office of the European Union: Luxemburg, 2011.
6. Klinglmair, M.; Sala, S.; Brandão, M. Assessing resource depletion in LCA: A review of methods and methodological issues. *Int. J. Life Cycle Assess.* **2014**, *19*, 580–592. [CrossRef]
7. Tilton, J.E. Exhaustible resources and sustainable development: Two different paradigms. *Resour. Policy* **1996**, *22*, 91–97. [CrossRef]
8. Sonnemann, G.; Gemechu, E.D.; Adibi, N.; de Bruille, V.; Bulle, C. From a critical review to a conceptual framework for integrating the criticality of resources into Life Cycle Sustainability Assessment. *J. Clean Prod.* **2015**, *94*, 20–34. [CrossRef]
9. Tilton, J.E. Depletion and the long-run availability of mineral commodities. In International Institute for Environment and Development, Prepared for Workshop on the Long-Run Availability of Mineral Commodities Sponsored by the Mining, Minerals and Sustainable Development Project and Resources for the Future. Washington, DC, USA, 22–23 April 2001. Available online: http://pubs.iied.org/pdfs/G01035.pdf (accessed on 8 July 2013).
10. Tilton, J.E. On borrowed time? In *Assessing the Threat of Mineral Depletion*; Resources for the Future: Washington, DC, USA, 2002.

11. Vieira, M.D.M.; Storm, P.; Goedkoop, M.J. Stakeholder Consultation: What do decision makers in public policy and industry want to know regarding abiotic resource use? In *Towards Life Cycle Sustainability Management*; Finkbeiner, M., Ed.; Springer: Berlin, Germany, 2011; pp. 27–34.

12. Drielsma, J.A.; Russel-Vaccari, A.J.; Drnek, T.; Brady, T.; Weihed, P.; Mistry, M.; Perez Simbor, L. Mineral resources in life cycle impact assessment—Defining the path forward. *Int. J. Life Cycle Assess.* **2016**, *21*, 85–105. [CrossRef]

13. Backman, C.-M. Global Supply and Demand of Metals in the Future. *J. Toxicol. Environ. Health A* **2008**, *71*, 1244–1253. [CrossRef] [PubMed]

14. Giurco, D.; Evans, G.; Cooper, C.; Mason, L.; Franks, D. Mineral Futures Discussion Paper: Sustainability Issues, Challenges and Opportunities. 2009. Available online: http://www.csiro.au/ Organisation-Structure/Flagships/Minerals-Down-Under-Flagship/mineral-futures/ mineral-futures-collaboration-cluster/minerals-futures-discussion-paper.aspx (accessed on 22 August 2012).

15. International Energy Agency. World Energy Outlook. International Energy Agency, 2011. Available online: http://www.iea.org/publications/freepublications/publication/WEO2011_WEB.pdf (accessed on 22 November 2013).

16. Meadows, D.; Randers, J.; Meadows, D. *Limits to Growth—The 30 Year Update*; Earthscan: London, UK, 2005.

17. Tilton, J.E.; Lagos, G. Assessing the long-run availability of copper. *Resour. Policy* **2007**, *32*, 19–23. [CrossRef]

18. Schneider, L.; Berger, M.; Schüler-Hainsch, E.; Knöfel, S.; Ruhland, K.; Mosig, J.; Bach, V.; Finkbeiner, M. The economic resource scarcity potential (ESP) for evaluating resource use based on life cycle assessment. *Int. J. Life Cycle Assess.* **2014**, *19*, 601–610. [CrossRef]

19. Ponsioen, T.C.; Vieira, M.D.M.; Goedkoop, M.J. Surplus cost as a life cycle impact indicator for fossil resource scarcity. *Int. J. Life Cycle Assess.* **2014**, *19*, 872–881. [CrossRef]

20. Goedkoop, M.; de Schryver, A. Mineral Resource Depletion. In ReCiPe 2008: A life cycle impact assessment method which comprises harmonised category indicators at the midpoint and the endpoint level. In *Report I: Characterisation Factors*, 1st ed.; Goedkoop, M., Heijungs, R., Huijbregts, M.A.J., de Schryver, A., Struijs, J., van Zelm, R., Eds.; Ministerie van Volkhuisvesting, Ruimtelijke Ordening en Milieubeheer: Den Haag, The Netherlands, 2009; pp. 108–117.

21. Dahlman, C.J. The problem of externality. *J. Law Econ.* **1979**, *22*, 141–162. [CrossRef]

22. Weidema, B.P. Using the budget constraint to monetarise impact assessment results. *Ecol. Econ.* **2009**, *68*, 1591–1598. [CrossRef]

23. Hunkeler, D., Lichtenvort, K., Rebitzer, G., Eds. *Environmental Life Cycle Costing*; Society of Environmental Toxicology and Chemistry (SETAC): New York, NY, USA, 2008.

24. Hochschorner, E.; Noring, M. Practitioners' use of life cycle costing with environmental costs—A Swedish study. *Int. J. Life Cycle Assess.* **2011**, *16*, 897–902. [CrossRef]

25. Vieira, M.D.M.; Goedkoop, M.J.; Storm, P.; Huijbregts, M.A.J. Ore Grade Decrease As Life Cycle Impact Indicator for Metal Scarcity: The Case of Copper. *Environ. Sci. Technol.* **2012**, *46*, 12772–12778. [CrossRef] [PubMed]

26. U.S. Geological Survey. *Mineral Commodity Summaries 2014*; U.S. Geological Survey: Reston, VA, USA, 2014. Available online: http://minerals.usgs.gov/minerals/pubs/mcs/2014/mcs2014.pdf (accessed on 3 October 2014).

27. UNEP. *Estimating Long-Run Geological Stocks of Metals*; UNEP International Panel on Sustainable Resource Management, Working Group on Geological Stocks of Metal: Paris, France, 2011. Available online: http://www.unep.org/resourcepanel/Portals/24102/PDFs/GeolResourcesWorkingpaperfinal040711.pdf (accessed on 24 December 2015).

28. Kelly, T.D.; Matos, G.R. Historical Statistics for Mineral and Material Commodities in the United States (2014 Version). Available online: http://minerals.usgs.gov/minerals/pubs/historical-statistics/ (accessed on 12 November 2014).

29. Nuclear Energy Agency and International Atomic Energy Agency (NEA-IAEA). *Uranium 2014: Resources, Production and Demand*, NEA No. 7209. Available online: http://www.oecd-nea.org/ndd/ pubs/2014/7209-uranium-2014.pdf (accessed on 12 November 2014).

30. Hall, S.; Coleman, M. *Critical Analysis of World Uranium Resources*; U.S. Scientific Investigations Report 2012–5239; U.S. Geological Survey: Reston, VA, USA, 2012.

31. Pincock, Allen and Holt. The Processing of Platinum Group Metals (PGM)—Part 1. Pincock Perspectives 89, March 2008. Available online: http://www.pincock.com/perspectives/Issue89-PGM-Processing-Part1.pdf (accessed on 5 November 2012).

32. Schneider, L.; Berger, M.; Finkbeiner, M. Abiotic resource depletion in LCA—Background and update of the anthropogenic stock extended abiotic depletion potential (AADP) model. *Int. J. Life Cycle Assess.* **2015**, *20*, 709–721. [CrossRef]

33. World Mine Cost Data Exchange. Cost Curves. Available online: http://www.minecost.com/curves.htm (accessed on 2 July 2014).

34. Bureau of Labor Statistics. CPI Inflation Calculator. Available online: http://www.bls.gov/data/inflation_calculator.htm (accessed on 9 July 2014).

35. R Development Core Team. R: A Language and Environment for Statistical Computing, Version 3.0.1 (2013–05–16). R Foundation for Statistical Computing: Vienna, Austria, 2013. Available online: http://www.R-project.org (accessed on 1 November 2013).

36. Coulas, M. Copper. In *Canadian Minerals Yearbook 2008*; Birchfield, G., Trelawny, P., Eds.; Natural Resources Canada: Vancouver, Canada, 2008. Available online: http://www.nrcan.gc.ca/mining-materials/markets/canadian-minerals-yearbook/2008/commodity-reviews/8522?destination=node/3012 (accessed on 3 September 2014).

37. Philips, W.G.B.; Edwards, D.P. Metal prices as a function of ore grade. *Resour. Policy* **1976**, *2*, 167–178. [CrossRef]

38. Steen, B. *A Systematic Approach to Environmental Priority Strategies in Product Development (EPS). Version 2000—Models and Data of the Default Method*; Chalmers University of Technology: Göteburg, Sweden, 1999.

39. Crowson, P. Mine size and the structure of costs. *Resour. Policy* **2003**, *29*, 15–36. [CrossRef]

40. Rudenno, V. *The Mining Valuation Handbook: Mining and Energy Valuation for Investors and Management*, 4th ed.; John Wiley & Sons Australia: Brisbane, Australia, 2012.

41. Swart, P.; Dewulf, J. Quantifying the impacts of primary metal resource use in life cycle assessment based on recent mining data. *Resour. Conserv. Recycl.* **2013**, *73*, 180–187. [CrossRef]

42. Gerst, M.D. Linking Material Flow Analysis and Resource Policy via Future Scenarios of In-Use Stock: An Example for Copper. *Environ. Sci. Technol.* **2009**, *43*, 6320–6325. [CrossRef] [PubMed]

43. Gerst, M.D. Revisiting the cumulative grade-tonnage relationship for major copper ore types. *Econ. Geol.* **2008**, *103*, 615–628. [CrossRef]

44. Reynolds, D.B. The mineral economy: How prices and costs can falsely signal decreasing scarcity. *Ecol. Econ.* **1999**, *31*, 155–166. [CrossRef]

45. Mudd, G. An analysis of historic production trends in Australian basic metal mining. *Ore Geol. Rev.* **2007**, *32*, 227–261. [CrossRef]

resources

MDPI

Article

Resource Efficiency Assessment—Comparing a Plug-In Hybrid with a Conventional Combustion Engine

Martin Henßler [1,*], Vanessa Bach [2,*], Markus Berger [2], Matthias Finkbeiner [2] and Klaus Ruhland [1]

[1] Group Environmental Protection, Daimler AG, Stuttgart 70546, Germany; klaus.ruhland@daimler.com
[2] Chair of Sustainable Engineering, Technische Universität Berlin, Straße des 17. Juni 135, Berlin 10623, Germany; markus.berger@tu-berlin.de (M.B.); matthias.finkbeiner@tu-berlin.de (M.F.)
* Correspondence: martin.henssler@daimler.com (M.H.), vanessa.bach@tu-berlin.de (V.B.); Tel.:+49-30-314-27941 (V.B.)

Academic Editor: Mario Schmidt
Received: 22 October 2015; Accepted: 22 December 2015; Published: 21 January 2016

Abstract: The strong economic growth in recent years has led to an intensive use of natural resources, which causes environmental stress as well as restrictions on the availability of resources. Therefore, a more efficient use of resources is necessary as an important contribution to sustainable development. The ESSENZ method presented in this article comprehensively assesses a product's resource efficiency by going beyond existing approaches and considering the pollution of the environment as well as the physical and socio-economic availability of resources. This paper contains a short description of the ESSENZ methodology as well as a case study of the Mercedes-Benz C-Class (W 205)—comparing the conventional C 250 (petrol engine) with the C 350 e Plug-In Hybrid (electric motor and petrol engine). By applying the ESSENZ method it can be shown that the use of more and different materials for the Plug-In-Hybrid influences the dimensions physical and socio-economic availability significantly. However, for environmental impacts, especially climate change and summer smog, clear advantages of the C 350 e occur due to lower demand of fossil energy carriers. As shown within the case study, the when applying the ESSENZ method a comprehensive evaluation of the used materials and fossil energy carriers can be achieved.

Keywords: resource efficiency; life cycle assessment; physical availability; socio-economic availability; environmental impact

1. Introduction

The demand of abiotic resources like metals, or fossil fuels has increased significantly in recent decades due to the global industrial and technological development. Additionally, the pollution of natural resources like water has risen as well. As these patterns of resource consumption will lead to an exceedance of the sustainable capacity of ecosystems worldwide, enhancing resource efficiency is a key goal of national and international strategies (e.g., [1,2]).

There is no commonly accepted definition of the term "resources" yet. Often it is defined as materials and energy as well as knowledge, services, or staff. With regard to the assessment of efficiency of a resource the term is mostly used evaluating the use of minerals, metals, and fossil energy carriers [2–5]. The Strategy on the sustainable use of natural resources [6] includes the environment in the definition as well, leading to a more comprehensive view of resources.

Resource efficiency is mostly regarded as a macro economic strategy because resources are key components of every society to sustain production and the wellbeing of current and future generations [5,7]. However, often resource efficiency is implemented on a micro economic level e.g., by reducing material inputs. Thus, measuring resource efficiency in a methodologically correct yet applicable way on a product level enables companies to address scarcity of resources and decrease environmental impacts on a corporate level.

Evaluating the resource efficiency of products can be established by the ESSENZ method (Integrated method to assess/measure resource efficiency) [8–11].

In the following sections, the applied ESSENZ method is introduced as well as the subjects of the case study. Results are shown for the individual dimensions as well as for the summarized results.

2. Method

In cooperation with TU Berlin (Chair of Sustainable Engineering), Daimler AG, Deutsches Kupferinstitut Berufsverband e. V., Evonik Industries AG, Siemens AG, ThyssenKrupp Steel Europe AG, and Wissenschaftlicher Gerätebau Dr.-Ing. Herbert Knauer GmbH a comprehensive method has been developed to measure resource efficiency of products. Overall, 18 categories and corresponding indicators were established to enable a holistic assessment of resource efficiency in the context of sustainable development considering the three dimensions "physical availability", "socio-economic availability", and "environmental impacts" (Figure 1) [8–11]. Within the ESSENZ approach next to the environment (including all environmental compartments e.g., water, air, soil), raw materials are considered as a resource as well. Even though the developed approach can be applied to all resources in theory, practical experience has so far been limited to metals, fossil energy carriers, as well as parts of the environment.

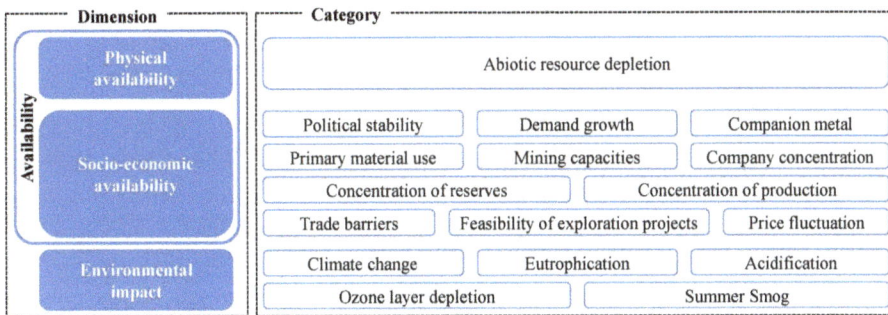

Figure 1. Dimensions and categories to assess resource efficiency within the ESSENZ method.

Existing geological deposits (physical availability) as well as socio-economic factors (socio-economic availability) might be restricting the availability of resources and thus influencing their supply security. The physical availability is evaluated by means of the abiotic depletion potential (ADP) indicator (baseline approach—ultimate reserves), which is subdivided to assess resource depletion of raw materials (ADP$_{elemental}$) and resource depletion of fossil fuels (ADP$_{fossil}$) [12]. To evaluate the socio-economic availability of resources, economic constraints leading to supply shortages along the product's value chain are quantified [13]. Possible constraints include, for example, the political stability of countries, which can be impaired due to factors such as corruption within the government, disrupting the capacity to effectively implement robust policies [14]. Based on existing work [15–17] 11 categories with corresponding category indicators are identified (Table 1). Characterization factors for all categories are determined based on the ecological scarcity approach [18,19]. Indicator values are compared with regard to a category specific target. These targets have been determined based on a stakeholder survey and expert interviews. Applying normalization and scaling (up to 1.5×10^{19} representing the overall production of all considered materials within the year 2013) the calculation of the final characterization factors is concluded. They are provided for a portfolio of 36 metals and four fossil energy carriers [8–11]. So far, the physical and socio-economic availability of a product can only be determined for the BoM of the considered product. Due to the fact that current LCA databases use economic allocation to assign metal contents from mixed ores to metal datasets, the mass of metals in the datasets does not reflect the physically present metal content but rather represents an over- or underestimation depending on the economic value of the considered metals.

Table 1. Socio-economic categories and related category indicators.

Category	Description	Category Indicator
Political stability	Governance stability of producing countries	World Governance Indicators [14]
Demand growth	Increase of demand over the last five years	Percentage of annual growth based on past developments (based on data from British Geological Service [20])
Companion metal	Companion metals within host metal ore bodies	Percentage of production as companion metal [21]
Primary material use	Recycling content of a material	Percentage of new material content [22]
Mining capacities	Overall mining time of a material considering current production	Reserve-to-annual-production ratio (based on data from United States Geological Service [23] and British Geological Service [20])
Company concentration	Company concentration based on producing companies	HHI[(1)]—index is calculated by squaring the market share of each company or country with regard to the production or reserves [24]
Concentration of reserves	Reserve concentration of certain materials based on reserves in countries	
Concentration of production	Concentration of mine production based on production in countries	
Trade barriers	Materials underlying trade barriers	Enabling Trade Index [25]
Feasibility of exploration projects	Political and societal factors influencing opening of mines	Policy Potential Index [26]
Price fluctuation	Unexpected price fluctuations	Volatility [27]

Note: [(1)] HHI: Herfindahl-Hirschman-Index.

For determining impacts of resource use on the environment five different indicators are applied using the CML-IA impact assessment method (CML 2001—Version: April 2013, baseline approach) [28] for the subjects' climate change, acidification, eutrophication, ozone layer depletion, and summer smog. The considered environmental impact categories as well as methods are chosen based on their applicability and maturity [29]. Thus, biodiversity and land use are not included as no adequate and applicable methods exist so far [30].

For an overall result regarding the resource efficiency of a product system the considered categories are analyzed together to achieve a comprehensive evaluation enabling meaningful decision making [8–11].

3. Case Study of Mercedes-Benz C-Class

In the following section, the resource efficiency assessment according to the ESSENZ method is presented for the example of the C-Class (W 205). The study compares the conventional C 250 (petrol engine) with the C 350 e Plug-In Hybrid (electric motor and petrol engine). First, in Section 3.1

technical data and a detailed analysis of materials for the considered passenger cars are provided. Results of the resource efficiency assessment with the ESSENZ method are shown in Section 3.2.

3.1. Product Documentation of the C 250 and C 350 e

This section documents significant specifications of the different variants of the C-Class analyzed in this study. Section 3.1.1. provides an overview of the technical data of the C 250 and C 350 e. The material composition is discussed in Section 3.1.2.

3.1.1. Technical Data

The Plug-In Hybrid model in the current C-Class, the C 350 e, combines a 60 kW electric motor and an externally rechargeable battery with a four-cylinder petrol engine with 155 kW (Table 2). The high voltage lithium-ion battery of the C 350 e has an energy content of 6.38 kWh. With the aid of the synchronous electric motor, the C 350 e has an all-electric range of 31 km. The certified combined consumption according to the New European Driving Circle (NEDC) of the C 350 e is 2.1 l and 11.0 kWh per 100 kilometer (ECE-R101). This corresponds to CO_2-emissions of 48 g/km.

Table 2. Technical data of C 250 and C 350 e [31].

Technical Data	C 250	C 350 e
Weight (kg)	1435 [1]	1705
Output (kW)	155	155 + 60 (electric motor)
Fuel consumption NEDC [2] combined (l/100 km)	5.3	2.1
Electric energy consumption NEDC [2] combined (kWh/100 km)	-	11.0
Electric range (km)	-	31
Driving share petrol engine (%)	100	45 [3]
Driving share electric motor (%)	-	55 [3]
CO_2 (g/km)	123	48

Notes: [1] Comparably equipped as C 350 e; [2] NEDC: New European Driving Circle; [3] Determination of electric driving share according to type approval directive ECE-R101; percentages related to driving distance.

The C 250 is powered by the 155 kW four-cylinder petrol engine (Table 2). The fuel consumption is 5.3 l/100 km (NEDC). This causes CO_2-emissions of 123 g/km which are more than twice as high as the CO_2-emissions of the C 350 e.

The C 250 and the C 350 e can be assumed to be functionally equivalent as they have similar driving performance as well as safety and comfort features (Table 2). The use phase is calculated on the basis of a mileage of 200,000 km. The key components of both vehicles (incl. battery) do not require replacement over the life cycle.

3.1.2. Material Composition

The weight and material data for the C 250 and C 350 e are determined on the basis of internal documentation of the components used in the vehicle (parts list, drawings). The kerb weight according to DIN 70020 (without driver and luggage, fuel tank 90 percent full) serves as a basis for the life cycle assessment (LCA). Figure 2 shows the material composition of the C 250 and C 350 e.

The weight of the C 250 is 1435 kg. The weight of the C 350 e is 1705 kg and thus 270 kg heavier than the C 250. Steel/ferrous materials account for slightly less than half the vehicle weight (approximately 47 percent) in both cars. The next largest shares are light alloys at 21 percent and polymer materials at 19 (C 250) and 21 percent (C 350 e). Service fluids and other metals comprise around 5 (C 350 e) and 6 percent (C 250); and 2 (C 250) and 5 percent (C 350 e), respectively. The proportions of other materials are somewhat lower, at about 4 percent.

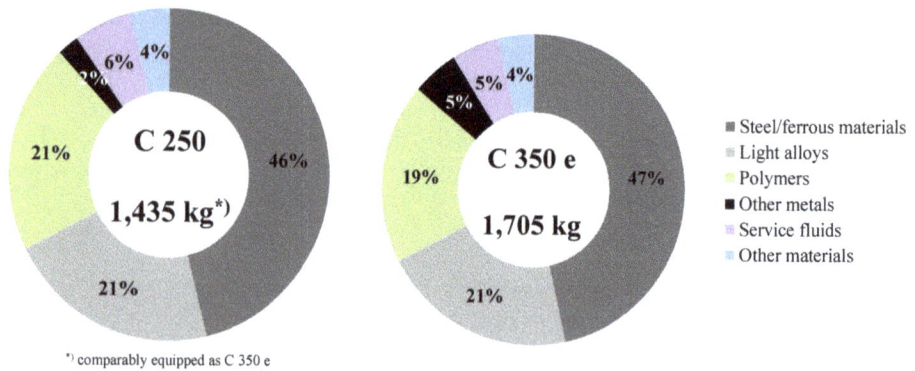

*) comparably equipped as C 350 e

Figure 2. Material composition of C 250 and C 350 e [31].

Figure 3 shows the main differences in weight and material mix in the modules exterior, interior, chassis, powertrain, and electric comparing the C 350 e with the C 250. The biggest difference can be found regarding electric constituents. Due to the hybrid components, especially the high voltage battery, the power electronics, and the cabling, the additional weight is about 140 kg. In the powertrain, an extra weight of about 76 kg is derived primarily from the electric motor. Larger breaks and tires as well as air suspension cause an extra weight of the chassis of about 66 kg. The weight of the exterior is about 25 kg higher due to the high voltage crash package of the battery. The alternative drive components and the related mix of materials—especially the material group of other metals—used in the C 350 e change the weight substantially compared to the conventional C-Class.

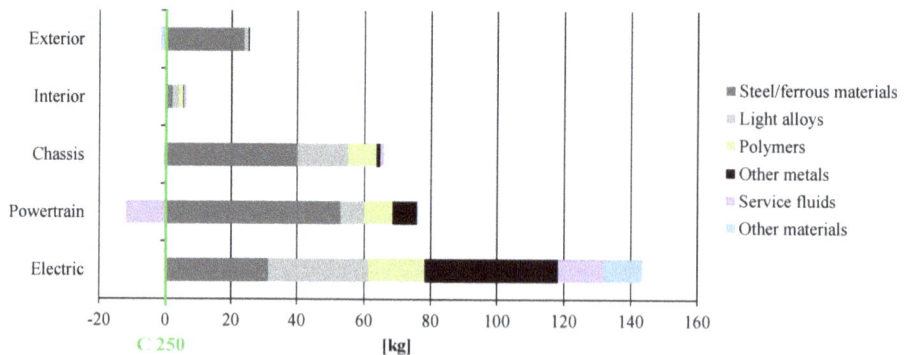

Figure 3. Main weight differences—C 350 e compared to the C 250.

3.2. Assessment of Different Resource Efficiency Dimensions Considered in the ESSENZ Method

The ESSENZ method described in Section 2 was applied, leading to the following results regarding physical availability (Section 3.2.1.), socio-economic availability (Section 3.2.2.), and environmental impacts (Section 3.2.3.). The section concludes by summarizing the results of all considered dimensions of the C 250 compared to the C 350 e (Section 3.2.4.).

For the C 350 e, two energy consumption scenarios for the use phase are considered. In addition to the EU electricity grid mix, electricity from hydro power is accounted for. The results of the use phase (electricity generation, fuel production, and operation) are based on the certified NEDC electricity/fuel consumption and the certified specific emissions of each car via a mileage of 200,000 km. The study includes environmental impacts of the recovery phase on the basis of the standard processes of drainage, shredding, and recovery of energy from the shredder light fraction. Environmental credits are not considered.

3.2.1. Physical Availability

Figure 4 shows the results for the category abiotic resource depletion (quantified by $ADP_{elemental}$ in kg Sb_{eq} and ADP_{fossil} in GJ). For the calculation of the $ADP_{elemental}$ only the BoM of the vehicle is taken into account since no consistent background data for the whole supply chain are available and no materials are required in the use phase. Assessing resource depletion of fossil energy carriers (ADP_{fossil}) the whole life cycle (car production, fuel production, operation, electricity generation, and end of life) of the vehicle is included. The additional hybrid-specific components (Section 3.1.2.) lead to a higher resource depletion potential of metals for the C 350 e, which is about 170 percent higher (0.63 kg Sb_{eq}) compared to the C 250 (0.23 kg Sb_{eq}).

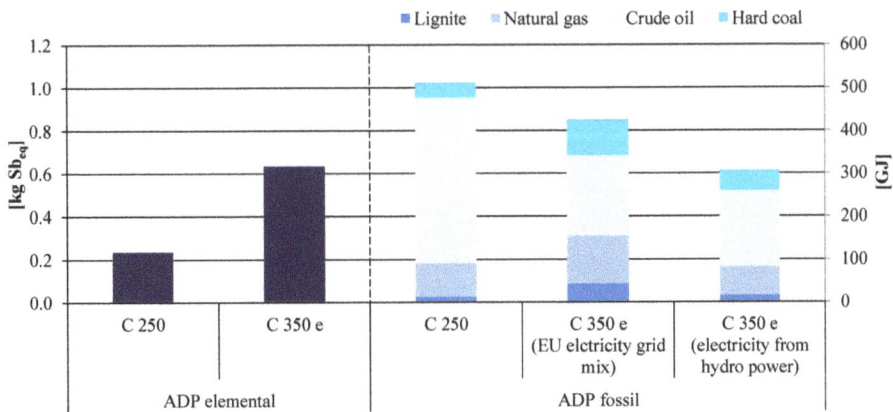

Figure 4. $ADP_{elemental}$ and ADP_{fossil}—C 250 compared to the C 350 e.

The abiotic depletion potential (ADP_{fossil}) of the C 350 e with electricity from hydro power is 40 percent lower (205.6 GJ) than of the C 250 (511 GJ—comprised of 385.6 GJ crude oil, 78.8 GJ natural gas, 34.6 GJ hard coal, and 12.7 GJ lignite). Due to additional hybrid-specific components in the car production and the generation of electricity during the operation phase, the consumption of natural gas, hard coal, and lignite rises for the C 350 e using EU electricity grid mix to 112.9 GJ (natural gas), 84.1 GJ (hard coal), and 41.8 GJ (lignite). Crude oil consumption can be reduced by over 50 percent to 185.6 GJ due to the high efficiency of the Plug-In Hybrid. When the vehicle is charged with renewably generated electricity, the consumption of lignite, natural gas, crude oil, and hard coal can be reduced further.

Evaluating the abiotic resource depletion enables a comprehensive assessment of the physical availability of metals and fossil energy carriers. Advantages in the production of conventional vehicle concepts compared to alternative vehicle concepts due to the reduced consumption of materials could be shown. Furthermore, alternative engines show benefits in the use phase (electricity generation, fuel production, and operation) due to lower fuel consumption.

3.2.2. Socio-Economic Availability

The dimension "socio-economic availability" is quantified by 11 categories (Section 2). Regarding the socio-economic availability of metals the BoMs of the respective cars are taken into account. As most fossil energy carriers are consumed in the use phase, the whole life cycle is considered when analyzing their socio-economic availability. The results of the comparison of C 250 and C 350 e are shown in Figure 5 (C 250 is scaled to 100 percent).

Overall, for the calculation of the socio-economic availability, 33 metals (Section 2) are taken into account in addition to the fossil energy carriers' lignite, natural gas, crude oil, and hard coal. As shown in Figure 5, the fossil energy carriers have little influence on the 11 categories. Exceptions are "Primary material use" (19 to 39 percent), "Price fluctuation" (12 to 28 percent), and "Company concentration" (2 to 5 percent). Due to the higher material consumption of the C 350 e the C 250 performs far better in all categories except for "Mining capacities". The source of electricity used to charge the C 350 e has no impact regarding the socio-economic availability. The C 350 e using EU electricity grid mix is almost on par with C 350 e using electricity from hydro power—except differences in the categories "Primary material use" and "Price fluctuation" due to higher amounts of used fossil energy carriers (see Section 3.2.1.).

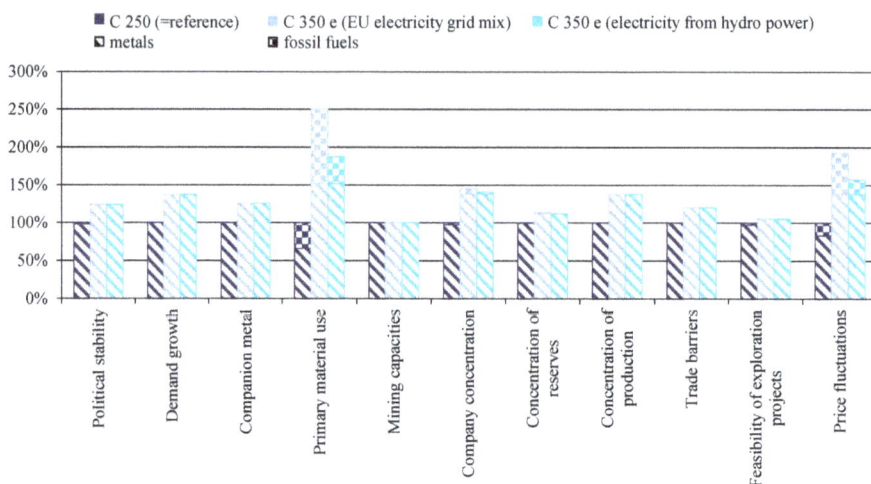

Figure 5. Assessment of the socio-economic availability—C 250 compared to the C 350 e.

The greatest differences between the C 250 and the C 350 e (EU electricity grid mix) occur in the categories "Primary material use" (+ 150 percent), "Price fluctuation" (+ 93 percent), "Company concentration" (+ 45 percent), "Concentration of production" (38 percent), and "Demand growth" (+ 37 percent). Regarding the categories "Demand growth" and "Primary material use", the differences are caused by the use of lithium, which is an essential part of the high voltage battery of the Plug-In Hybrid. The categories "Company concentration", "Concentration of production", and "Price fluctuation" are primarily affected by rare earth elements, which are mainly required for the magnets of the electric motor.

For the categories "Political stability", "Companion metal", "Mining capacities", "Concentration of reserves", "Trade barriers", and "Feasibility of exploration projects" the C 350 e performs up to 26 percent worse than the C 250. These categories are mainly affected by platinum and palladium (exhaust catalyst), magnesium (alloy material), lithium (high voltage battery), rare earth elements (electric motor), and tantalum (condensers).

Overall, it can be concluded that the socio-economic categories in this case study are particularly affected by platinum, palladium, magnesium, lithium, rare earth and tantalum. Except for magnesium, all materials occur in the car in very small amounts.

3.2.3. Environmental Impacts

Figure 6 shows the results of the considered environmental categories climate change (CO_{2eq}-emissions), eutrophication (phosphate$_{eq}$-emissions), acidification (SO_{2eq}-emissions), and summer smog (ethene$_{eq}$-emissions) over the individual life cycle phases (car production, fuel production, operation, electricity generation, and end of life) of the C 250 and C 350 e.

The production of the C 350 e entails visibly higher CO_{2eq}-emissions on account of the additional hybrid-specific components. The CO_{2eq}-emissions in the production phase (11.3 t CO_{2eq}) are 32 percent higher than those of the C 250 (8.6 t CO_{2eq}). Over the entire life cycle the Plug-In Hybrid has clear advantages as external charging with the EU electricity grid mix can cut overall CO_{2eq}-emissions by about 13 percent (4.9 t CO_{2eq}) compared to the C 250. A reduction of 39 percent (15.3 t CO_{2eq}) is possible through the use of renewably generated electricity from hydro power.

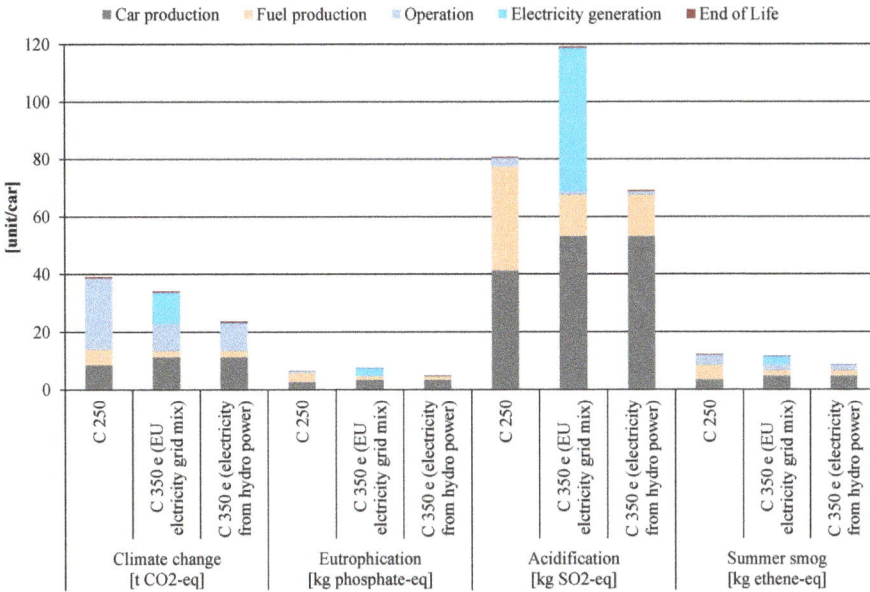

Figure 6. Selected environmental categories—C 250 compared to the C 350 e (unit/car) [24].

Considering the category eutrophication, the C 350 e using electricity from hydro power causes the lowest emissions with 4.9 kg phosphate$_{eq}$. Charging the C 350 e with EU electricity grid mix the phosphate$_{eq}$-emissions add up to 7.5 kg phosphate$_{eq}$, thus being respectively 15 and 53 percent higher than the phosphate$_{eq}$-emissions of the C 250 (6.5 kg phosphate$_{eq}$) or the C 350 e with electricity from hydro power.

In the production phase of the C 350 e (53.2 kg SO_{2eq}) the SO_{2eq}-emissions are 29 percent higher than those of the C 250 (41.2 kg SO_{2eq}). In the use phase (electricity generation, fuel production, and operation) most SO_{2eq}-emissions (65.5 kg SO_{2eq}) are produced during the charging of the vehicle with EU electricity grid mix. Thus, 40 and 76 percent more emissions occur than for the C 250 or the C 350 e with electricity from hydro power respectivly. Over the entire life cycle, the C 350 e with electricity from hydro power saves 42 percent (50.0 kg SO_{2eq}) in comparison to the C 350 e with EU electricity grid mix and 14 percent (11.6 kg SO_{2eq}) compared to the C 250.

The summer smog emissions during production of the C 350 e (4.7 kg ethene$_{eq}$) are 36 percent higher than those of the C 250 (3.4 kg ethene$_{eq}$). Regarding the use phase (electricity generation, fuel production and operation) summer smog emissions can be reduced by 22 percent charging the C 350 e with EU electricity grid mix respectively by 55 percent using electricity from hydro power compared to the C 250. The highest summer smog emissions with 12.1 kg ethene$_{eq}$ are caused by the C 250.

Compared to the C 350 e charged with EU electricity grid mix as well as renewable generated electricity summer smog can be reduced by 5 or 29 percent, translating in a reduction of 11.5 or 8.6 kg ethene$_{eq}$.

In conclusion, it can be stated that over the entire life cycle the C 350 e using electricity from hydro power has clear benefits in all considered categories (shown in Figure 6) compared to the C 250. If the EU electricity grid mix is used for charging advantages with respect to climate change and summer smog occur. However, with regard to eutrophication and acidification, the C 350 e respectively has 15 percent (1.0 kg ethene$_{eq}$) and 48 percent (38.4 kg SO$_{2eq}$) more impacts than the C 250.

3.2.4. Summary of the Results

Figure 7 shows the summary of the three dimensions considered in this case study. The reference C 250 is scaled to 100 percent.

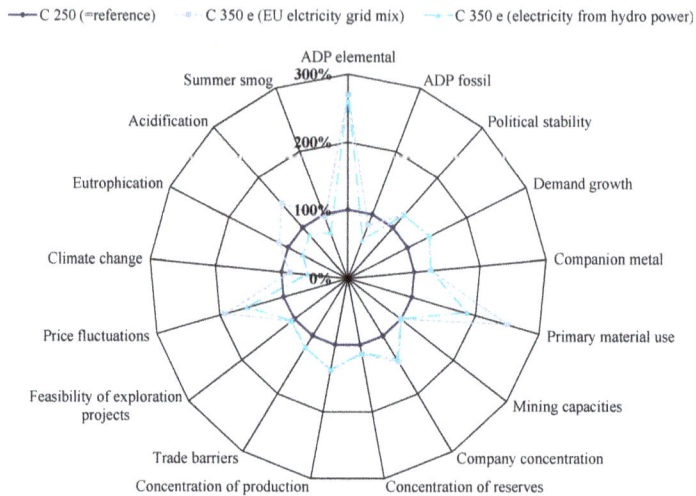

Figure 7. Summary of resource efficiency dimensions of the ESSENZ method—C 250 compared to the C 350 e.

It can be seen that the C 250 performs better within the category abiotic resource depletion of metals (ADP$_{elemental}$) compared to the C 350 e due to its lower overall use of metals. This is also reflected in the dimension "socio economic availability". As the C 250 has a higher fossil energy carrier consumption in the use phase, it performs worse in the category resource depletion of fossil energy carriers (ADP$_{fossil}$). Both C 350 e's have advantages in the use phase (electricity generation, fuel production and operation) due to their lower fuel consumption.

Only minor differences for the dimension "socio-economic availability" as well as the category resource depletion of raw materials (ADP$_{elemental}$) occur when comparing the C 350 e charged with electricity from hydro power with the C 350 e charged with EU electricity grid mix. The metals used for car production influence the categories more significantly than the energy consumption in the use phase. For the category resource depletion of fossil energy carriers (ADP$_{fossil}$) the C 350 e (electricity from hydro power) performs only slightly better.

The results of the environmental impacts measured over the entire life cycle show clear advantages for the C 350 e especially for the categories climate change and summer smog, regardless of the kind of electricity used for external charging of the Plug-In Hybrid. The results for the categories eutrophication and acidification, however, depend on the electricity used for charging the vehicle. Using EU electricity grid mix eutrophying or acidifying emissions of C 350 e are higher than those of C 250. Using electricity from hydro power, the eutrophication potential or acidification potential are lower compared to C 250.

4. Conclusions

The ESSENZ method allows a transparent evaluation of product systems with regard to the physical and socio-economic availability of fossil energy carriers and metals as well as related environmental impacts over the life cycle.

The case study presented in this article—comparing a plug-in hybrid with a conventional engine—is a good example of why such a comprehensive assessment is necessary. The higher use of material resources for the Plug-In Hybrid vehicle has a strong influence on socio-economic and physical availability, whereas some of the environmental impact categories (e.g., climate change and summer smog) show clear advantages for the C 350 e due to the lower demand of fossil energy carriers. As a consequence, it is necessary for a scientifically robust and sustainability oriented resource assessment to consider both materials and energy resources as well as the whole lifecycle of the product.

The comprehensive ESSENZ method enables the user to transparently evaluate various views on the multitude of parameters applicable. It further empowers companies to take appropriate actions regarding the specific materials used in their products, e.g., material specific sourcing strategies or development of recycling technologies.

Acknowledgments: The paper is based on the project "ESSENZ—Integrierte Methode zur ganzheitlichen Berechnung/Messung von Ressourceneffizienz" which was funded by the German Federal Ministry of Education and Research (BMBF)—project code number: 033R094A. We would like to thank the Federal Ministry of Education and Research for their financial support.

Author Contributions: Martin Henßler was the lead in carrying out the case study. Vanessa Bach was leading in the development of the ESSENZ method. Markus Berger and Matthias Finkbeiner both provided substantial contributions to the design of the ESSENZ method as well as valuable comments to deepen the conclusion. Klaus Ruhland gave substantial input to the design of the case study as well as valuable comments to draw the conclusion. All authors proofread and approved the final manuscript.

Conflicts of Interest: The authors declare no conflict of interest.

References

1. European Commission. *Roadmap to a Resource Efficient Europe*; European Commission: Brüssel, Belgium, 2011.
2. Bundesregierung Deutschland. Nationale Nachhaltigkeitsstrategie Fortschrittsbericht 2012. Available online: http://www.bundesregierung.de/Webs/Breg/DE/Themen/Nachhaltigkeitsstrategie/1-die-nationale-na chhaltigkeitsstrategie/nachhaltigkeitsstrategie/_node.html;jsessionid=AB7764D74BA79942AF3B8D3300348 11D.s3t2 (accessed on 18 January 2016).
3. Ritthoff, M.; Rohn, H.; Liedtke, C. *Calculating MIPS—Resource Productivity of Products and Services*; Environment and Energy Report; Wuppertal Institute for Climate: Wuppertal, Germany, 2002. Available online: http://epub.wupperinst.org/files/1577/WS27e.pdf (accessed on 18 January 2016).
4. European Commission. Resource Efficiency. The Roadmap's Approach to Resource Efficiency Indicators. Available online: http://ec.europa.eu/environment/resource_efficiency/targets_indicators/roadmap/ index_en.htm (accessed on 18 December 2015).
5. Schneider, L.; Bach, V.; Finkbeiner, M. LCA Perspectives for Resource Efficiency Assessment. In *Special Types of LCA*; Springer: Berlin, Germany, 2015.
6. Thematic Strategy on the Sustainable Use of Natural Resources. Available online: http://eur-lex.europa.eu/ legal-content/EN/TXT/?uri=celex%3A52005DC0670 (accessed on 18 January 2016).
7. Mikesell, R.F. Viewpoint—Sustainable development and mineral resources. *Resour. Policy* **1994**, *20*, 83–86. [CrossRef]
8. Bach, V.; Berger, M.; Henßler, M.; Kirchner, M.; Leiser, S.; Mohr, L.; Rother, E.; Ruhland, K.; Schneider, L.; Tikana, L.; *et al. Integrierte Methode zur Ganzheitlichen Berechnung/Messung von Ressourceneffizienz (ESSENZ-Methode)*; Springer, 2016; in press.

9. Bach, V.; Schneider, L.; Berger, M.; Finkbeiner, M. ESSENZ-Projekt: Entwicklung einer Methode zur Bewertung von Ressourceneffizienz auf Produktebene. In *3. Symposium Rohstoffeffizienz und Rohstoffinnovation*; Fraunhofer Verlag: Stuttgart, Germany, 2014; pp. 463–474.

10. Bach, V.; Schneider, L.; Berger, M.; Finkbeiner, M. Methoden und Indikatoren zur Messung von Ressourceneffizienz im Kontext der Nachhaltigkeit. In *Recycling und Rohstoffe*; Thome-Kozmiensky, K.J., Goldmann, D., Eds.; TK Verlag: Neuruppin, Germany, 2014; pp. 87–101.

11. Bach, V.; Berger, M.; Helbig, T.; Finkbeiner, M. Measuring a product's resource efficiency—A case study of smartphones. In Proceedings of the CILCA 2015—VI International Conference on Life Cycle Assessment, Lima, Peru, 13–16 March 2015.

12. Van Oers, L.; de Koning, A.; Guinée, J.B.; Huppes, G. Abiotic Ressource Depletion in LCA Improving Characterisation Factors for Abiotic Resource Depletion as Recommended in the Dutch LCA Handbook, 2002. Available online: http://www.leidenuniv.nl/cml/ssp/projects/lca2/report_abiotic_depletion_web.pdf (accessed on 24 December 2015).

13. Schneider, L.; Berger, M.; Schüler-Hainsch, E.; Knöfel, S.; Ruhland, K.; Mosig, J.; Bach, V.; Finkbeiner, M. The economic resource scarcity potential (ESP) for evaluating resource use based on life cycle assessment. *Int. J. Life Cycle Assess.* **2014**, *19*, 601–610. [CrossRef]

14. The World Bank Group. The Worldwide Governance Indicators. Available online: http://info.worldbank.org/governance/wgi/index.aspx#home (accessed on 18 December 2015).

15. Graedel, T.E.; Barr, R.; Chandler, C.; Chase, T.; Choi, J.; Christoffersen, L.; Friedlander, E.; Henly, C.; Jun, C.; Nassar, N.T.; et al. Methodology of metal criticality determination. *Environ. Sci. Technol.* **2012**, *46*, 1063–1070. [CrossRef] [PubMed]

16. Schneider, L. A Comprehensive Approach to Model Abiotic Resource Provision Capability in the Context of Sustainable Development. Ph.D. Thesis, Technische Universität Berlin, Berlin, Germany, 6 August 2014.

17. Erdmann, L.; Behrendt, S.; Feil, M. Kritische Rohstoffe für Deutschland "Identifikation aus Sicht Deutscher Unternehmen Wirtschaftlich Bedeutsamer Mineralischer Rohstoffe, Deren Versorgungslage Sich Mittel-Bis Langfristig als Kritisch Erweisen Könnte", 2011. Available online: https://www.kfw.de/Download-Center/Konzernthemen/Research/PDF-Dokumente-Sonderpublikationen/Kritische-Rohstoffe-KF.pdf (accessed on 24 December 2015).

18. Frischknecht, R.; Steiner, R.; Jungbluth, N. *The Ecological Scarcity Method—Eco-Factors 2006*; Federal Office for the Environment FOEN: Bern, Switzerland, 2009.

19. Müller-Wenk, R.; Ahbe, S.; Braunschweig, A.; Müller-Wenk, R. *Methodik für Ökobilanzen auf der Basis ökologischer Optimierung*; Schriftenreihe Umwelt 133; Bundesamt für Umwelt BAFU (former BUWAL): Bern, Switzerland, 1990.

20. Brown, J.; Wrighto, C.E.; Raycraft, E.R.; Shaw, R.A.; Deady, E.A.; Rippingale, J.; Bide, T.; Idoine, N. *World Mineral Production*; British Geological Survey: Keyworth, UK, 2014.

21. Angerer, G.; Erdmann, L.; Marscheider-Weidemann, F.; Scharp, M.; Lüllmann, A.; Handke, V.; Marwerde, M. *Rohstoffe für Zukunftstechnologien Rohstoffe für Zukunftstechnologien*; Fraunhofer IRB Verlag: Stuttgart, Germany, 2009. Available online: http://www.isi.fraunhofer.de/isi-wAssets/docs/n/de/publikationen/Schlussbericht_lang_20090515_final.pdf (accessed on 18 January 2016).

22. Graedel, T.E. *UNEP Recycling Rates of Metals—A Status Report*; Working Group on the Global Metal Flows; International Resource Panel, United Nations Environment Programme: Nairobi, Kenya, 2011. Available online: http://www.unep.org/resourcepanel-old/portals/24102/pdfs/UNEP_report2_Recycling_130920.pdf (accessed on 24 December 2015).

23. United States. Geological. Service (USGS). Commodity Statistics and Information, 2015. Available online: http://minerals.usgs.gov/minerals/pubs/commodity/ (accessed on 20 May 2004).

24. Rhoades, S.A. The Herfindahl-Hirschman index. *Fed. Reserv. Bull.* **1993**, *79*, 188.

25. Hanouz, M.D.; Geiger, T.; Doherty, S. *The Global Enabling Trade Report 2014*; World Economic Forum: Geneva, Switzerland, 2014.

26. Cervantes, M.; McMahon, F.; Wilson, A. *Survey of Mining Companies: 2012/2013*; Fraser Institut: Vancouver, BC, Canada, 2013.

27. Bundesanstalt für Geowissenschaften und Rohstoffe. In *Volatilitätsmonitor*; Bundesanstalt für Geowissenschaften und Rohstoffe: Hanover, Germany, 2014.

28. Guinée, J.B.; Gorrée, M.; Heijungs, R.; Huppes, G.; Kleijn, R.; de Koning, A.; van Oers, L.; Sleeswijk, A.W.; Suh, S.; de Haes, H.A.U.; *et al. Handbook on Life Cycle Assessment Operational Guide to the ISO Standards*; Kluwer Academic Publishers: New York, NY, USA, 2002.
29. Lehmann, A.; Bach, V.; Finkbeiner, M. Product environmental footprint in policy and market decisions—Applicability and impact assessment. *Integr. Environ. Assess. Manag.* **2015**. [CrossRef] [PubMed]
30. Finkbeiner, M.; Ackermann, R.; Bach, V.; Berger, M.; Brankatschk, G.; Chang, Y.-J.; Grinberg, M.; Lehmann, A.; Martínez-Blanco, J.; Minkov, N.; *et al.* Challenges in Life Cycle Assessment: An Overview of Current Gaps and Research Needs. In *LCA Compendium—The Complete World of Life Cycle Assessment—Volume 1: Background and Future Prospects in Life Cycle Assessment*; Klöpffer, W., Ed.; Springer: Dodrecht, The Netherlands, 2014; pp. 207–258.
31. Daimler, A.G. Life cycle—Environmental Certificate Mercedes-Benz C-Class including Plug-In Hybrid C 350 e. *Stuttgart* **2015**, in press.

resources

MDPI

Article

Dynamic Ecocentric Assessment Combining Emergy and Data Envelopment Analysis: Application to Wind Farms

Mario Martín-Gamboa [†] **and Diego Iribarren** [†,*]

Systems Analysis Unit, Instituto IMDEA Energía, Av. Ramón de la Sagra 3, Móstoles E-28935, Spain; mario.martin@imdea.org
* Correspondence: diego.iribarren@imdea.org; Tel.: +34-91-737-11-19
† These authors contributed equally to this work.

Academic Editor: Mario Schmidt
Received: 3 December 2015; Accepted: 21 January 2016; Published: 29 January 2016

Abstract: Most of current life-cycle approaches show an anthropocentric standpoint for the evaluation of human-dominated activities. However, this perspective is insufficient when it comes to assessing the contribution of natural resources to production processes. In this respect, emergy analysis evaluates human-driven systems from a donor-side perspective, accounting for the environmental effort performed to make the resources available. This article presents a novel methodological framework, which combines emergy analysis and dynamic Data Envelopment Analysis (DEA) for the ecocentric assessment of multiple resembling entities over an extended period of time. The use of this approach is shown through a case study of wind energy farms. Furthermore, the results obtained are compared with those of previous studies from two different angles. On the one hand, a comparison with results from anthropocentric approaches (combined life cycle assessment and DEA) is drawn. On the other hand, results from similar ecocentric approaches, but without a dynamic model, are also subject to comparison. The combined use of emergy analysis and dynamic DEA is found to be a valid methodological framework for the computation of resource efficiency and the valuation of ecosystem services. It complements traditional anthropocentric assessments while appropriately including relevant time effects.

Keywords: dynamic data envelopment analysis; efficiency; emergy; life cycle assessment

---

1. Introduction

The current level of extraction of primary resources—around 60 billion tons per year globally—is not sustainable [1]. Moreover, the global consumption of natural resources for the production of goods and services is expected to increase significantly in next decades [1,2]. Fundamental changes and policy measures are therefore required in order to promote a shift in production processes, supply-chain management and consumption patterns [2–4]. In this sense, a dematerialization strategy should be followed, *i.e.*, decoupling natural resource use from economic growth and social prosperity [1,2]. This calls for a sustainable and efficient use of resources, security in the supply of raw materials and reduction in life-cycle environmental impacts. Furthermore, within this context, novel methodological frameworks for assessing systems performance are needed to ensure the rational use of natural resources [4].

Life-cycle methodologies are one of the main instruments for the comprehensive evaluation of supply-chains and product systems due to its holistic nature [5]. However, most of current life-cycle approaches show an anthropocentric standpoint for the evaluation of human-dominated activities, generally disregarding flows beyond the present state of a resource in the current ecosphere [6,7].

Moreover, the evaluation of resource criticality within the life cycle assessment (LCA) methodology remains underdeveloped [8]. Hence, LCA alone is insufficient when it comes to assessing flawlessly the contribution of natural resources to product systems. In this respect, emergy analysis evaluates human-driven systems from a donor-side perspective, accounting for the environmental effort performed to make the resources available [9,10].

Emergy estimates the solar energy previously provided to generate a product [11,12]. It is interpreted as the memory of the geobiosphere exergy provision related to economic systems through the use of natural resources [13,14]. Thus, emergy analysis complements LCA by providing a donor-side perspective, a unified measure of the provision of environmental support, and an indication of the work of the environment that would be needed to replace what is consumed [9,10]. LCA and emergy analysis have been applied jointly to a number of case studies, e.g., within the primary [15], building [16] and energy [6,7,17] sectors.

When assessing multiple homogeneous entities involving a number of inputs and outputs, another suitable tool for combination with emergy analysis is data envelopment analysis (DEA), which is a linear programming methodology to measure the relative efficiency of multiple "decision making units" (DMUs, *i.e.*, the homogeneous entities under assessment) [18]. In this respect, Iribarren *et al.* [10] proposed the combined use of emergy and DEA—the Em + DEA method—to enhance benchmarking processes in terms of environmental sustainability, offering an ecocentric life-cycle perspective.

In case of systems with significant time variability, the use of the Em + DEA method—and, in general, of the currently available life-cycle approaches coupled with DEA [19]—provides a static picture of the "historical" use of natural resources. To deal with this flaw, dynamic (time-dependent) DEA models might be used [20]. This article presents a novel methodological framework combining emergy analysis and dynamic DEA—the Em + DynDEA method hereinafter—for the ecocentric assessment of multiple resembling entities over an extended period of time. The feasibility of this novel approach is shown through a case study of wind farms. Furthermore, the results obtained are compared with those of previous studies from two different angles. On the one hand, a comparison with results from anthropocentric approaches (combined LCA and DEA) is drawn. On the other hand, results from similar ecocentric approaches, but without a dynamic model, are also subject to comparison.

2. Material and Methods

2.1. Em + DynDEA Method

The goal of the study is to extend the applicability of the available Em + DEA method [10] to time-dependent systems by using a dynamic DEA model. The resultant Em + DynDEA method is outlined in Figure 1. Three key steps are included: (i) data collection, (ii) emergy analysis, and (iii) dynamic DEA.

The first step focuses on the preparation of the life-cycle inventory (LCI) of each individual DMU for each individual period. In the second step, these LCIs are used to carry out the emergy analysis of each DMU for each period by using the SCALE software developed by Marvuglia *et al.* [21]. In this study, the total value of emergy (in seMJ, *i.e.*, million solar emjoules) is considered, which is calculated as the sum of the seven resource categories computed with SCALE (*viz.*, fossil resources, metal ores, mineral resources, nuclear resources, renewable energy resources, water resources, and land resources). It should be noted that the use of the SCALE software involves an appropriate consideration of the emergy algebra rules [21], thus distinguishing itself from other attempts that define solar energy demand indicators based on the use of characterization factors suitable for implementation into LCA software [22].

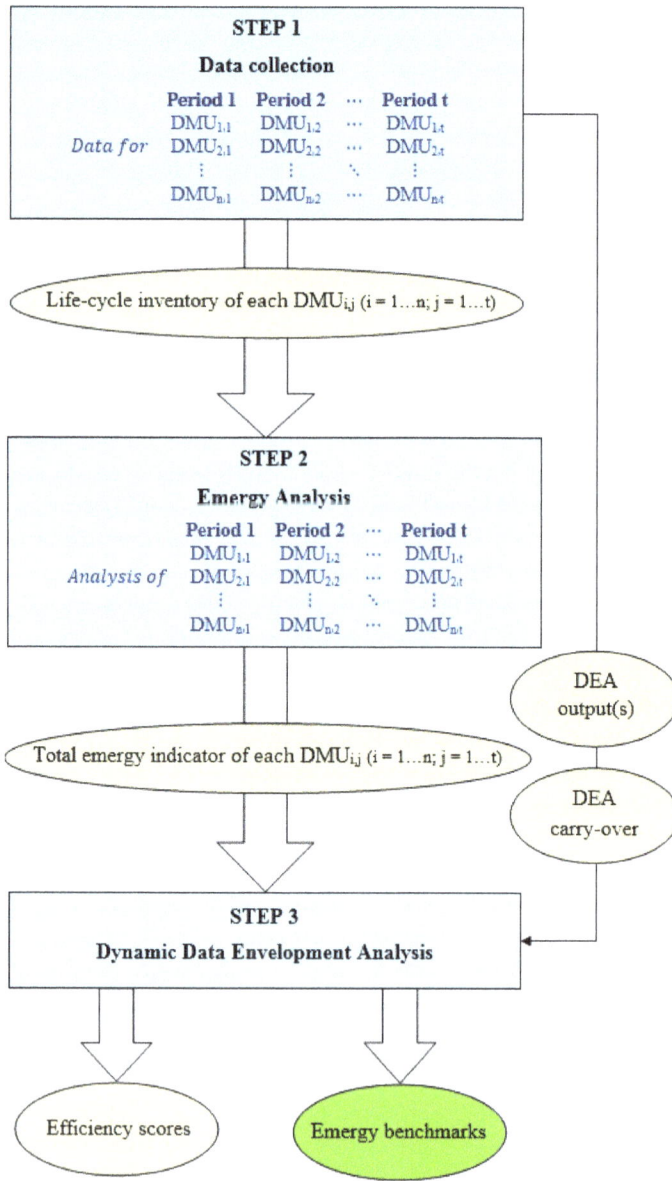

Figure 1. Representation of the Em + DynDEA method.

The final step of the Em + DynDEA method provides relative efficiency scores (Φ) and emergy benchmarks (*i.e.*, target values for efficient performance) through dynamic DEA of the whole sample of *n* DMUs over *t* periods. At each period *j*, each DMU *i* has its respective inputs and outputs along with a carry-over (link) that connects two consecutive periods. In the Em + DynDEA method, the total emergy value is used as the DEA input. The DEA matrix is completed by including the output(s) to which the inputs refer, as well as the carry-over that links the next period *j+1*. Carry-overs are usually classified into four categories: desirable (good) link; undesirable (bad) link; discretionary (free) link;

and non-discretionary (fixed) link [20]. Once the DEA matrix is ready, key features must be selected regarding the DEA model, which includes the type of carry-over, the orientation of the model and the display of the production possibility set [10]. When compared to other alternatives for efficiency computation, the use of DEA benefits from advantages mainly associated with the definition of the production possibility set and, subsequently, of the efficient frontier, making DEA also a powerful methodology for the benchmarking of performance indicators [18,19]. When using dynamic DEA under time-dependent situations, additional benefits are associated with the use of time-series data and the carry-over [20]. The need for both inventory data and carry-over data may give rise to potential links to other analytical methods such as (time-dependent) material flow analysis [23].

When solving the dynamic DEA model, efficiency scores are provided for each DMU and period ($\Phi_{i,j}$), in addition to the overall efficiency of each DMU (Φ_i, average of the t efficiency scores of DMU i). The relative efficiency scores lead to distinguish between efficient ($\Phi = 1$) and inefficient ($\Phi < 1$) entities within a certain number of periods. Moreover, target emergy values are provided as the desired benchmarks to turn inefficient entities into efficient DMUs.

2.2. Definition of the Case Study

The feasibility of the Em + DynDEA method is shown through a case study of wind farms (WFs). Wind power is selected as case study because it shows time variability and it is a key power generation option in a wide range of countries [24,25]. A sample of 16 WFs located in Spain (regions of Castile-La Mancha and Andalusia) is used in the current study as homogeneous entities (DMUs). The analysis is carried out on an annual basis for a time frame of five years, from 2007 to 2011, a key period for wind power growth in Spain [26].

Table 1 presents the evolution of wind power generation over 2007–2011 for the evaluated sample. These data are retrieved from the annual statistics published in the Spanish Ministry of Industry, Energy and Tourism database [27]. During this period, the average production of the whole set of WFs was 85.55 GWh·y^{-1}. Additionally, the installed capacity and the average wind speed ranged from 30 to 50 MW and from 5.5 to 10.5 m·s^{-1}, respectively. The installed capacity (also reported in Table 1) was unintentionally constant for the specific sample under study.

Table 1. Electricity produced by the wind farms over 2007–2011 (GWh) and installed capacity (MW).

Wind Farm Code	Year 2007	Year 2008	Year 2009	Year 2010	Year 2011	Installed Capacity
WF1	91.88	102.01	91.63	101.93	109.25	44.80
WF2	81.60	93.36	91.26	82.27	89.98	33.40
WF3	109.29	128.76	124.62	123.29	107.22	50.00
WF4	93.30	100.39	99.85	96.61	83.92	49.50
WF5	102.29	113.00	99.91	100.86	91.71	49.50
WF6	98.51	94.41	95.71	94.78	83.84	49.30
WF7	22.09	90.26	76.75	73.30	79.89	48.00
WF8	89.61	100.89	98.65	94.08	80.98	45.05
WF9	83.22	92.03	93.98	88.01	74.30	41.65
WF10	76.41	84.91	71.06	73.98	77.45	42.00
WF11	77.98	88.84	86.26	83.80	73.67	37.60
WF12	82.21	90.83	90.19	84.01	77.83	37.40
WF13	86.89	91.71	85.48	81.41	84.69	36.96
WF14	62.77	71.00	70.68	69.44	68.54	31.45
WF15	60.76	65.04	62.64	61.15	53.63	31.02
WF16	66.30	61.83	74.35	71.23	59.74	30.00

Figure 2 illustrates the structure of the dynamic DEA study of WFs. This analysis involves 16 WFs ($n = 16$) over 5 years ($t = 5$). For each of the years, each WF encompasses one input (total emergy) and one output (electricity production). The choice of total emergy as an input provides an exhaustive evaluation of actual resource use, integrating all the information on primary material consumption into a single input. This consumption is associated mainly with the central element of a WF, the wind turbine, which includes four main components (concrete foundations, tower, nacelle, and rotor) and a large number of subcomponents and electrical constituents [28]. The preparation of the LCI of each WF—as required for the emergy analysis—includes all these items. Further details about the definition of wind farms as DMUs can be found in Iribarren *et al.* [10,28].

Finally, a link between years is used to connect the activity of the WFs and take into account efficiency changes. The choice of the carry-over(s) ultimately depends on the needs of the analyst for the specific case study addressed. Examples of different types of appropriate links reflecting actual characteristics of carry-over activities can be found in the scientific literature [20]. For the case study of wind farms, as shown in Figure 2, generation capacity is selected as a discretionary (free) carry-over. This type of free link has an indirect effect on the efficiency score due to the continuity condition between two consecutive periods, implicit condition in the formulation of the dynamic DEA model [20]. Since the value of the free link can be increased or decreased from the observed one, the link deviation from the current value is calculated as a slack variable [20].

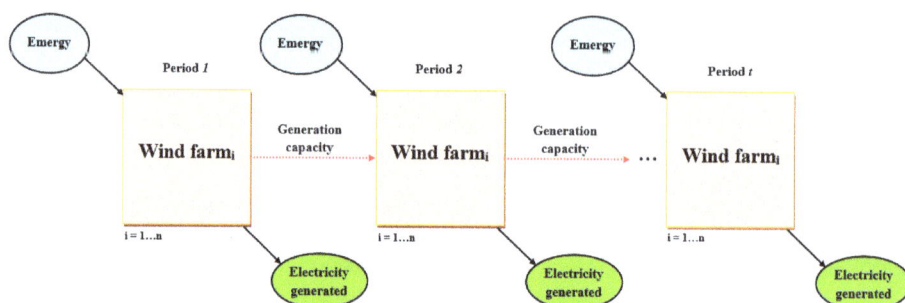

Figure 2. Key components of the dynamic DEA study of wind farms.

3. Results and Discussion

3.1. Application of the Em + DynDEA Method

Table 2 presents the main average inventory data for the set of DMUs over the period 2007–2011. The structure of the inventory is in accordance with previous studies [10,28]. This average inventory shows the convenience of carrying out individual analyses in order to avoid misleading result interpretation owing to relevant standard deviations. These deviations are due to both time and DMU variations. The use of the average inventory in Table 2 is therefore avoided [10]. Instead, the individual LCI of each WF for each period is used according to the proposed Em + DynDEA method. Foreground data acquisition is based on Iribarren *et al.* [28], while inventory data for background processes are taken from the ecoinvent database [29] and air emissions from diesel combustion are estimated using data from the European Environment Agency (EEA) [30].

Each individual LCI is implemented into SimaPro (PRé Consultants, Amersfoort, The Netherlands) [31] and, subsequently, SCALE [21] in order to perform emergy computation. The total emergy value of each WF for each period is obtained by adding the corresponding emergy results of the seven resource categories [21]. The total emergy values are then implemented into the DEA matrix (Table 3) as a DEA input.

Table 2. Average inventory data of the sample of wind farms (values per MWh of generated power).

Inputs	Units	Average ± Standard Deviation
From Nature		
Land	m^2a	4.24 ± 1.60
Kinetic energy (converted)	MWh	1.08 ± 0.00
From the Technosphere		
Concrete (foundations)	kg	19.30 ± 11.63
Iron (foundations)	kg	0.69 ± 0.42
Steel (foundations)	kg	0.41 ± 0.25
Steel (tower)	kg	3.33 ± 1.82
Paint (tower)	g	44.97 ± 24.6
Iron (nacelle)	kg	0.51 ± 0.31
Steel (nacelle)	kg	0.49 ± 0.31
Silica (nacelle)	g	9.49 ± 5.71
Copper (nacelle)	kg	0.10 ± 0.06
Plastic (nacelle)	g	13.79 ± 8.30
Aluminum (nacelle)	g	4.96 ± 2.99
Fiberglass (nacelle)	g	22.06 ± 13.29
Epoxy resin (nacelle)	g	33.08 ± 19.92
Epoxy resin (rotor)	kg	0.31 ± 0.33
Fiberglass (rotor)	kg	0.15 ± 0.07
Iron (rotor)	kg	0.38 ± 0.23
Lubricating oil	g	16.85 ± 6.70
Diesel, combusted	g	44.15 ± 44.17
Outputs	**Units**	**Average ± Standard Deviation**
Products		
Electricity	MWh	1.00 ± 0.00
Waste to treatment		
Waste to recycling	kg	6.60 ± 3.83
Waste to incineration	kg	0.47 ± 0.23

Table 3. DEA matrix for the Em + DynDEA study (values per MWh of generated power).

Wind Farm	Year 2007		Year 2008		Year 2009		Year 2010		Year 2011	
Code	EI	CCO	EI	CCO	EI	CCO	EI	CCO	EI	CCO
WF1	1.54×10^8	0.49	1.39×10^8	0.44	1.55×10^8	0.49	1.40×10^8	0.44	1.31×10^8	0.41
WF2	2.00×10^8	0.41	1.75×10^8	0.36	1.79×10^8	0.37	1.98×10^8	0.41	1.82×10^8	0.37
WF3	1.18×10^8	0.46	1.01×10^8	0.39	1.05×10^8	0.40	1.06×10^8	0.41	1.21×10^8	0.47
WF4	3.29×10^8	0.53	3.06×10^8	0.49	3.07×10^8	0.50	3.17×10^8	0.51	3.65×10^8	0.59
WF5	1.72×10^8	0.48	1.57×10^8	0.44	1.76×10^8	0.50	1.75×10^8	0.49	1.92×10^8	0.54
WF6	2.50×10^8	0.50	2.60×10^8	0.52	2.57×10^8	0.52	2.59×10^8	0.52	2.92×10^8	0.59
WF7	1.20×10^9	2.17	2.97×10^8	0.53	3.48×10^8	0.63	3.64×10^8	0.65	3.35×10^8	0.60
WF8	2.51×10^8	0.50	2.23×10^8	0.45	2.28×10^8	0.46	2.39×10^8	0.48	2.77×10^8	0.56
WF9	2.50×10^8	0.50	2.27×10^8	0.45	2.22×10^8	0.44	2.37×10^8	0.47	2.79×10^8	0.56
WF10	1.42×10^8	0.55	1.28×10^8	0.49	1.52×10^8	0.59	1.47×10^8	0.57	1.40×10^8	0.54
WF11	2.99×10^8	0.48	2.63×10^8	0.42	2.71×10^8	0.44	2.78×10^8	0.45	3.16×10^8	0.51
WF12	2.26×10^8	0.45	2.06×10^8	0.41	2.07×10^8	0.41	2.22×10^8	0.45	2.39×10^8	0.48
WF13	2.64×10^8	0.43	2.51×10^8	0.40	2.68×10^8	0.43	2.82×10^8	0.45	2.71×10^8	0.44
WF14	2.56×10^8	0.50	2.27×10^8	0.44	2.28×10^8	0.44	2.32×10^8	0.45	2.35×10^8	0.46
WF15	3.16×10^8	0.51	2.96×10^8	0.48	3.07×10^8	0.50	3.14×10^8	0.51	3.57×10^8	0.58
WF16	1.26×10^8	0.45	1.34×10^8	0.49	1.13×10^8	0.40	1.17×10^8	0.42	1.39×10^8	0.50

Notes: EI: emergy input ($seMJ \cdot y^{-1} \cdot MWh^{-1}$); CCO: capacity carry-over ($kW \cdot MWh^{-1}$).

The DEA matrix presented in Table 3 undergoes dynamic DEA. The selected model is a dynamic input-oriented slacks-based measure of efficiency model with constant returns to scale (DSBM-I-CRS model) [20]. This model is an extension of the slacks-based measure framework proposed by Tone [32] and Pastor *et al.* [33]. The choice of the key features of the model (non-radial metrics, input orientation and constant returns to scale) is based on previous DEA studies of wind farms [10,28]. Thus, the study focuses on the reduction of the emergy input while maintaining the observed output level. This is in line with the final goal of minimizing resource consumption. Finally, regarding the selected carry-over, the stability over time observed for the generation capacity supports its use as a carry-over, performing the continuity condition between consecutive periods in the dynamic model [20]. The suitability of generation capacity as a discretionary (free) link is in agreement with other DEA studies in the scientific literature [34].

The relative efficiency of each WF for each period, as well as the overall relative efficiency of each WF, is calculated by implementing the DEA matrix in DEA-Solver Pro (Saitech, Holmdel, NJ, USA) [35]. The same relevance is considered for the five individual periods (2007, 2008, 2009, 2010 and 2011). Table 4 shows the individual and overall efficiency scores of the assessed WFs. According to the overall scores, only 1 (WF3) out of 16 WFs is found to be efficient through the entire time frame (*i.e.*, overall Φ = 100%). In terms of individual periods, 2011 is—on average—the year with higher efficiency, arising as the only period with two efficient entities: WF1 and WF3. Regarding inefficient entities, more than half of the WFs present scores below 50%. Although the calculation of emergy benchmarks (*i.e.*, target emergy values) is feasible, it is out of the scope of the present study.

Table 4. Individual and overall efficiency scores (%) of the sample of wind farms.

DMU Code	Φ Year 2007	Φ Year 2008	Φ Year 2009	Φ Year 2010	Φ Year 2011	Overall Φ
WF1	81.85	82.01	81.95	81.86	100.00	85.83
WF2	59.28	59.50	59.44	59.44	71.38	61.81
WF3	100.00	100.00	100.00	100.00	100.00	100.00
WF4	36.05	36.24	36.18	36.16	36.08	36.14
WF5	68.68	68.92	68.86	68.87	68.85	68.84
WF6	47.46	47.69	47.58	47.58	47.58	47.58
WF7	34.30	34.14	34.03	34.02	37.48	34.79
WF8	47.22	47.43	47.37	47.35	47.28	47.33
WF9	47.36	47.58	47.52	47.50	47.43	47.48
WF10	83.33	83.56	83.61	83.59	87.85	84.39
WF11	39.65	39.85	39.79	39.78	39.65	39.74
WF12	52.30	52.53	52.46	52.44	52.32	52.41
WF13	44.84	45.06	44.95	44.95	44.83	44.93
WF14	46.26	46.48	46.42	46.41	54.89	48.09
WF15	37.47	37.66	37.59	37.58	37.52	37.56
WF16	94.21	94.82	94.27	94.34	94.55	94.44

3.2. Comparison of Ecocentric and Anthropocentric Approaches

A comparison between the efficiency scores calculated for a given period through the ecocentric Em + DynDEA method and those calculated through the anthropocentric LCA + DEA method for the same sample of wind farms is carried out in this section. The LCA + DEA efficiency scores are computed for the reference year 2010 as explained in Iribarren *et al.* [28], but using the reduced sample of WFs considered in the present article. These LCA + DEA efficiency scores are calculated through the five-step LCA + DEA method [28] and, therefore, characterize the operational performance of the evaluated WFs following an anthropocentric perspective [19]. Because the comparison is limited to efficiency scores, thus excluding the benchmarking of operational and environmental targets, the choice of the life cycle impact assessment method does not affect the results presented in this section.

Figure 3 shows the comparison between Em + DynDEA and LCA + DEA scores for the year 2010. As it can be observed, 3 entities (WF2, WF3 and WF16) are found to operate efficiently when using the anthropocentric (and static) approach, whereas the ecocentric (and dynamic) approach leads to identify only one efficient entity (WF3). In particular, WF2 constitutes a singular case since it is deemed efficient according to the LCA + DEA method, but significantly inefficient (Φ = 0.59) according to the Em + DynDEA method. Moreover, 75% of the WFs present efficiency scores above 0.6 when using the anthropocentric approach, while more than half of the WFs present scores below 0.5 when using the ecocentric method. According to other authors [20], the overestimation of efficiency scores in separate models (*i.e.*, models with no linkage between consecutive periods) is a common result since these models measure each term efficiency as completely independent, while dynamic models provide stable scores reflecting the continuity of links between periods. Hence, differences between Em + DynDEA and LCA + DEA results are associated with both the ecocentric/anthropocentric nature of the inputs (emergy *versus* operational inputs) and the dynamic/static character of the method. Nevertheless, it should be noted that Em + DynDEA and LCA + DEA are not seen as irreconcilable methods but as complementary approaches to enrich decision-making processes.

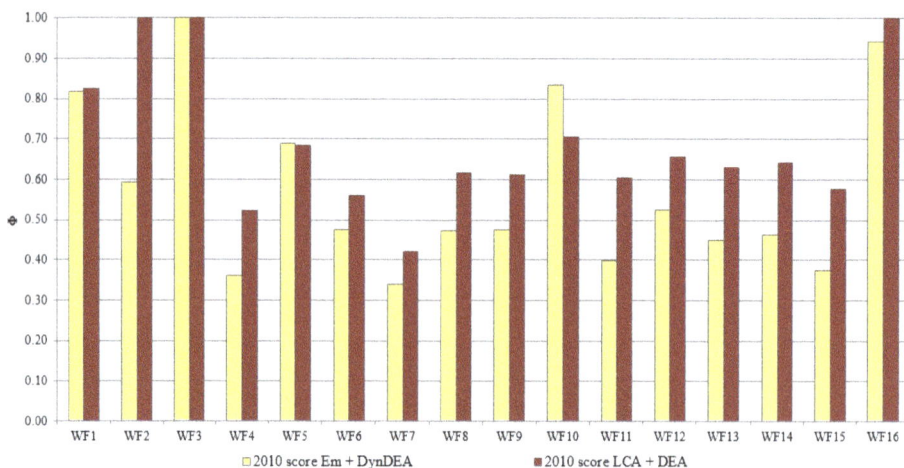

Figure 3. Comparison of the efficiency scores from ecocentric and anthropocentric approaches.

3.3. Comparison of Dynamic and Static Ecocentric Approaches

A comparison between the efficiency scores calculated for a given period through the dynamic Em + DynDEA method and those calculated through the static Em + DEA method for the same sample of wind farms is also carried out. The Em + DEA efficiency scores are computed for the reference year (2010) as detailed in Iribarren *et al.* [10], but using the reduced sample of WFs considered in the present work.

Figure 4 shows the comparison between the Em + DynDEA and Em + DEA scores for the year 2010. As it can be observed in this figure, the use of the static approach leads to identify WF2 and WF3 as efficient entities, while the dynamic approach leads to only one efficient DMU (WF3). Even though the efficiency scores are generally similar for both ecocentric approaches, the choice of a static/dynamic approach is likely to affect the identification of efficient entities (e.g., WF2) as well as the ranking of DMUs (e.g., WF10). Because emergy values are used as inputs in both approaches, the singularities found in the comparison are associated with the inclusion of the carry-over [20] and the different structure of the DEA matrix [10]. In this sense, the Em + DynDEA method should be understood as an enhancement of the Em + DEA method when evaluating entities with significant time variability.

Resources **2016**, *5*, 8

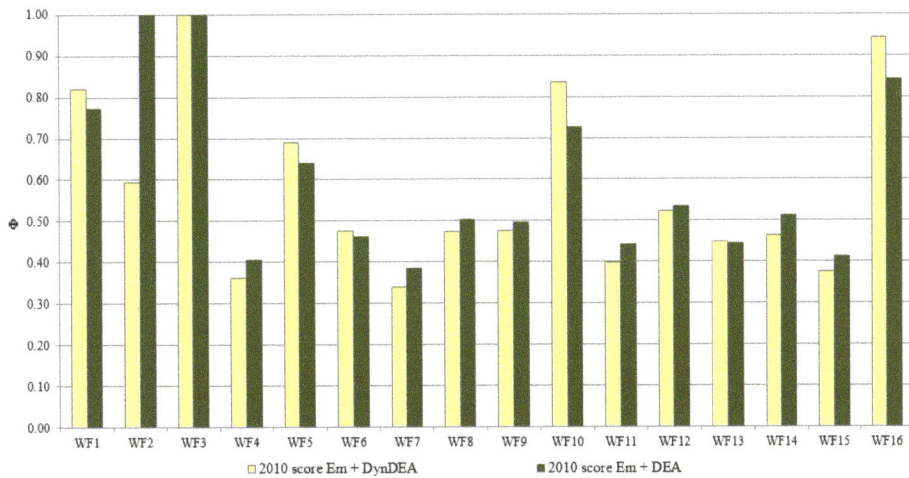

Figure 4. Comparison of the efficiency scores from dynamic and static ecocentric approaches.

4. Conclusions

When dealing with multiple similar entities, the use of emergy analysis as a life-cycle approach in combination with DEA facilitates a thorough evaluation and benchmarking of the environmental effort made by the biosphere to replace the resources consumed. In particular, when the entities under assessment involve relevant time variations, the combination of emergy analysis with dynamic DEA is concluded to be feasible. Under these circumstances, the use of this novel dynamic ecocentric approach—the Em + DynDEA method—is recommended, rather than the use of static ecocentric approaches. This should be understood as a helpful complement to anthropocentric approaches in order to enrich decision-making processes. As shown through the case study of wind farms, the combined use of emergy analysis and dynamic DEA succeeds in computing resource efficiency for the valuation of ecosystem services, thus complementing anthropocentric assessments while appropriately including relevant time effects.

Acknowledgments: All work presented in this article is part of the internal research activity at the Instituto IMDEA Energía (Spain).

Author Contributions: Both Mario Martín-Gamboa and Diego Iribarren contributed equally to the development of the Em + DynDEA method and its application to wind farms.

Conflicts of Interest: The authors declare no conflict of interest.

References

1. Bliesner, A.; Liedtke, C.; Welfens, M.J.; Baedeker, C.; Hasselkuß, M.; Rohn, H. "Norm-oriented interpretation learning" and resource use: The concept of "open-didactic exploration" as a contribution to raising awareness of a responsible resource use. *Resources* **2014**, *3*, 1–30. [CrossRef]
2. Liedtke, C.; Bienge, K.; Wiesen, K.; Teubler, J.; Greiff, K.; Lettenmeier, M.; Rohn, H. Resource use in the production and consumption system—The MIPS approach. *Resources* **2014**, *3*, 544–574. [CrossRef]
3. Bahn-Walkowiak, B.; Steger, S. Resource targets in Europe and worldwide: An overview. *Resources* **2015**, *4*, 597–620. [CrossRef]
4. Huysman, S.; Sala, S.; Mancini, L.; Ardente, F.; Alvarenga, R.A.F.; De Meester, S.; Mathieux, F.; Dewulf, J. Toward a systematized framework for resource efficiency indicators. *Resour. Conserv. Recy.* **2015**, *95*, 68–76. [CrossRef]

5.   Hagelaar, G.J.L.F.; van der Vorst, J.G.A.J. Environmental supply chain management: Using life cycle assessment to structure supply chains. *Int. Food Agribus. Manage. Rev.* **2001**, *4*, 399–412. [CrossRef]

6.   Buonocore, E.; Vanoli, L.; Carotenuto, A.; Ulgiati, S. Integrating life cycle assessment and emergy synthesis for the evaluation of a dry steam geothermal power plan in Italy. *Energy* **2015**, *86*, 476–487. [CrossRef]

7.   Neri, E.; Rugani, B.; Benetto, E.; Bastianoni, S. Emergy evaluation *vs.* life cycle-based embodied energy (solar, tidal and geothermal) of wood biomass resources. *Ecol. Indic.* **2014**, *36*, 419–430. [CrossRef]

8.   Sonnemann, G.; Gemechu, E.D.; Adibi, N.; De Bruille, V.; Bulle, C. From a critical review to a conceptual framework for integrating the criticality of resources into life cycle sustainability assessment. *J. Clean Prod.* **2015**, *94*, 20–34. [CrossRef]

9.   Raugei, M.; Rugani, B.; Benetto, E.; Ingwersen, W.W. Integrating emergy into LCA: Potential added value and lingering obstacles. *Ecol. Model.* **2014**, *271*, 4–9. [CrossRef]

10. Iribarren, D.; Vázquez-Rowe, I.; Rugani, B.; Benetto, E. On the feasibility of using emergy analysis as a source of benchmarking criteria through data envelopment analysis: A case study for wind energy. *Energy* **2014**, *67*, 527–537. [CrossRef]

11. Odum, H.T. Self-organization, transformity, and information. *Science* **1988**, *242*, 1132–1139. [CrossRef] [PubMed]

12. Odum, H.T. *Environmental Accounting: Emergy and Environmental Decision Making*; John Wiley & Sons: New York, NY, USA, 1996.

13. Sciubba, E.; Ulgiati, S. Emergy and exergy analyses: Complementary methods or irreducible ideological options? *Energy* **2005**, *30*, 1953–1988. [CrossRef]

14. Raugei, M. Emergy indicators applied to human economic systems—A word of caution. *Ecol. Model.* **2011**, *222*, 3821–3822. [CrossRef]

15. Pizzigallo, A.C.I.; Granai, C.; Borsa, S. The joint use of LCA and emergy evaluation for the analysis of two Italian wine farms. *J. Environ. Manage.* **2008**, *86*, 396–406. [CrossRef] [PubMed]

16. Reza, B.; Sadiq, R.; Hewage, K. Emergy-based life cycle assessment (Em-LCA) of multi-unit and single-family residential buildings in Canada. *Int. J. Sust. Built Environ.* **2014**, *3*, 207–224. [CrossRef]

17. Brown, M.T.; Raugei, M.; Ulgiati, S. On boundaries and 'investments' in emergy synthesis and LCA: A case study on thermal *vs.* photovoltaic electricity. *Ecol. Indic.* **2012**, *15*, 227–235. [CrossRef]

18. Cooper, W.W.; Seiford, L.M.; Tone, K. *Data Envelopment Analysis: A Comprehensive Text with Models, Applications, References and DEA-Solver Software*; Springer: New York, NY, USA, 2007.

19. Vázquez-Rowe, I.; Iribarren, D. Review of life-cycle approaches coupled with data envelopment analysis: Launching the CFP + DEA method for energy policy making. *Sci. World J.* **2015**, *2015*. [CrossRef] [PubMed]

20. Tone, K.; Tsutsui, M. Dynamic DEA: A slacks-based measure approach. *Omega* **2010**, *38*, 145–156. [CrossRef]

21. Marvuglia, A.; Benetto, E.; Rios, G.; Rugani, B. SCALE: Software for calculating emergy based on life cycle inventories. *Ecol. Model.* **2013**, *248*, 80–91. [CrossRef]

22. Rugani, B.; Huijbregts, M.A.J.; Mutel, C.; Bastianoni, S.; Hellweg, S. Solar energy demand (SED) of commodity life cycles. *Environ. Sci. Technol.* **2011**, *45*, 5426–5433. [CrossRef] [PubMed]

23. Zhou, Y.; Yang, N.; Hu, S. Industrial metabolism of PVC in China: A dynamic material flow analysis. *Resour. Conserv. Recy.* **2013**, *73*, 33–40. [CrossRef]

24. Kumar, Y.; Ringenberg, J.; Depuru, S.S.; Devabhaktuni, V.K.; Lee, J.W.; Nikolaidis, E.; Andersen, B.; Afjeh, A. Wind energy: Trends and enabling technologies. *Renew. Sust. Energ. Rev.* **2016**, *53*, 209–224. [CrossRef]

25. Kaplan, Y.A. Overview of wind energy in the world and assessment of current wind energy policies in Turkey. *Renew. Sust. Energ. Rev.* **2015**, *43*, 562–568. [CrossRef]

26. *Macroeconomic Study on the Impact of Wind Energy in Spain*; Spanish Wind Energy Association; AEE: Madrid, Spain, 2011.

27. Spanish Ministry of Industry, Energy and Tourism. Available online: http://www.minetur.gob.es (accessed on 3 December 2015).

28. Iribarren, D.; Martín-Gamboa, M.; Dufour, J. Environmental benchmarking of wind farms according to their operational performance. *Energy* **2013**, *61*, 589–597. [CrossRef]

29. Frischknecht, R.; Jungbluth, N.; Althaus, H.J.; Doka, G.; Heck, T.; Hellweg, S.; Hischier, R.; Nemecek, T.; Rebitzer, G.; Spielmann, M.; *et al. Overview and Methodology*; Ecoinvent Report No. 1; Swiss Centre for Life Cycle Inventories: Dübendorf, Switzerland, 2007.

30. European Environment Agency. *EMEP/EEA Air Pollutant Emission Inventory Guide Book 2009*; EEA: Copenhagen, Denmark, 2009.

31. Goedkoop, M.; Oele, M.; Leijting, J.; Ponsioen, T.; Meijer, E. *Introduction to LCA with SimaPro*; PRé Consultants: Amersfoort, the Netherlands, 2013.

32. Tone, K. A slacks-based measure of efficiency in data envelopment analysis. *Eur. J. Oper. Res.* **2001**, *130*, 498–509. [CrossRef]

33. Pastor, J.T.; Ruiz, J.L.; Sirvent, I. An enhanced DEA Russell graph efficiency measure. *Eur. J. Oper. Res.* **1999**, *115*, 596–607. [CrossRef]

34. Tone, K.; Tsutsui, M. Dynamic DEA with network structure: A slacks-based measure approach. *Omega* **2014**, *42*, 124–131. [CrossRef]

35. Saitech, Data Envelopment Analysis Software. Available online: http://www.saitech-inc.com/Products/Prod-DSP.asp (accessed on 3 December 2015).

resources

MDPI

Article

Evaluation of Abiotic Resource LCIA Methods

Rodrigo A. F. Alvarenga [1,*], **Ittana de Oliveira Lins** [2,†] **and José Adolfo de Almeida Neto** [2,†]

[1] Departamento de Engenharia Ambiental, Programa de Pós-graduação em Ciências Ambientais, Universidade do Estado de Santa Catarina (UDESC), Av. Luiz de Camões, 2090 Conta Dinheiro, Lages-SC 88.520-000, Brazil

[2] Departamento de Ciências Agrárias e Ambientais, Programa de Pós-graduação em Desenvolvimento e Meio Ambiente, Universidade Estadual de Santa Cruz (UESC), Rod. Ilhéus-Itabuna, km 16 s/n-Salobrinho-Ilhéus-BA 45.662-900, Brazil; Ittanalins@gmail.com (I.O.L.); jalmeida@uesc.br (J.A.A.N.)

* Correspondence: alvarenga.raf@gmail.com; Tel.: +55-49-3289-9295

† These authors contributed equally to this work.

Academic Editor: Mario Schmidt
Received: 9 December 2015; Accepted: 17 February 2016; Published: 29 February 2016

Abstract: In a life cycle assessment (LCA), the impacts on resources are evaluated at the area of protection (AoP) with the same name, through life cycle impact assessment (LCIA) methods. There are different LCIA methods available in literature that assesses abiotic resources, and the goal of this study was to propose recommendations for that impact category. We evaluated 19 different LCIA methods, through two criteria (scientific robustness and scope), divided into three assessment levels, *i.e.*, resource accounting methods (RAM), midpoint, and endpoint. In order to support the assessment, we applied some LCIA methods to a case study of ethylene production. For RAM, the most suitable LCIA method was CEENE (Cumulative Exergy Extraction from the Natural Environment) (but SED (Solar Energy Demand) and ICEC (Industrial Cumulative Exergy Consumption)/ECEC (Ecological Cumulative Exergy Consumption) may also be recommended), while the midpoint level was ADP (Abiotic Depletion Potential), and the endpoint level was both the Recipe Endpoint and EPS2000 (Environmental Priority Strategies). We could notice that the assessment for the AoP Resources is not yet well established in the LCA community, since new LCIA methods (with different approaches) and assessment frameworks are showing up, and this trend may continue in the future.

Keywords: abiotic; resource; life cycle assessment; LCA; life cycle impact assessment; LCIA; method; Brazil

1. Introduction

Natural resources are essential for our society, either for provision, supporting, regulating or cultural services [1]. However, due to the world population growth, together with the increase in the consumption per capita and poor resource management, we are being led to a sustainability crisis. Natural resources may be classified in several ways, (a) renewable or non-renewable, (b) stocks, funds or flows, (c) biotic or abiotic, among other classifications [2]. Regarding the latter, biotic resources are those that come from living organisms, while abiotic resources are the result of past biological processes (e.g., crude oil) or chemical processes (e.g., metal).

One of the tools that may assist in sustainable resource management, at the industrial scale, is a Life Cycle Assessment (LCA). Resources are seen in two ways in an LCA: (1) In one way, they are the inputs needed in industrial processes for the production of a product, and in this sense they are evaluated at the life cycle inventory (LCI) stage; and (2) in another way, they are evaluated as an area of protection (AoP), in life cycle impact assessments (LCIAs), *i.e.*, natural resources are one type of environmental impact assessed, and there are different methods to evaluate these impacts.

According to traditional classifications [2,3], these methods may be categorized into three groups: (1) Resource accounting methods (RAM), which make a more simplified impact analysis, focused mainly on grouping the resources into single score indicators, as energy or mass; (2) Midpoint resource depletion methods, that go beyond RAM, evaluating impacts related to resource depletion due to its use, as the use-to-availability ratio; and (3) Endpoint resource depletion methods, that go even beyond the previous group, taking into account the consequences of resource depletion, in many cases, through backup technology [4,5], e.g., evaluating the extra effort (energy or cost) needed to extract less economically feasible resources.

Due to the lack of consensus in the LCA community regarding impacts on natural resources, there are other approaches of how the LCIA methods evaluate that AoP. Rorbech *et al.* [6] classified LCIA methods into three groups: (1) Methods that account for the consumption of limited resources, which rather evaluate the resource competition and assume that they are exchangeable (as RAM); (2) Methods that evaluate the depletion of resources, which may be subdivided into midpoint and endpoint (and according to the authors would better represent the AoP Resources); and (3) Methods that evaluate the extra effort needed in the future due to actual resource extraction (e.g., Recipe Endpoint), which according to the authors do not represent a specific AoP as Resources, but are midpoint impacts that affect other AoPs (human health and natural environment). Dewulf *et al.* [7] suggested new AoPs for LCA and Life Cycle Sustainability Assessment (LCSA), proposing five perspectives: (P1) the safeguard subject is the resource itself; (P2) the concern is the capacity of this resource to generate provisional services; (P3) the safeguard subject is the capacity of this resource to generate other ecosystem services; (P4) where consequential aspects are considered, e.g., socioeconomic mechanisms are taken into account; and (P5) the concern is human well-being, giving a rather holistic perception by grouping all previous perspectives. The authors also mention that P4 and P5 go beyond classical LCA, and would better fit in LCSA.

Even though there are different ways of grouping LCIA methods in LCA, and how they affect (different) AoPs, in this manuscript we used the rather traditional overview, proposed by ILCD (International Reference Life Cycle Data System) and Swart *et al.* [2,3] (Figure 1). In this sense, there are some studies that already critically evaluated different LCIA methods, for instance, Liao *et al.* [8] evaluated thermodynamic-based RAM, and pointed out the Cumulative Exergy Extraction from the Natural Environment (CEENE) [9] and the Solar Energy Demand (SED) [10], as the recommended LCIA methods for that approach. In ILCD [3], where different LCIAs were evaluated in order to make recommendations for the European context, the Abiotic Depletion Potential (ADP) [11], adapted to the reserve base (Reserve base, according to the USGS (United States Geological Survey), accounts for all reserves that have the actual potential of extraction and may be economically viable in the future. The ADP method originally considered the ultimate reserve, which would be the amount of a certain resource in the Earth's crust), was recommended for midpoint assessment, while there was no recommendation for endpoint LCIA methods. The Life Cycle Initiative, from UNEP-SETAC (United Nations Environment Programme and the Society for Environmental Toxicology and Chemistry), is an ongoing project to create a worldwide consensus on recommendations of LCIA methods (http://www.lifecycleinitiative.org/activities/phase-i/life-cycle-impact-assessment-programme/) [12] which may be considered as a step forward to what has been done by ILCD [3].

Due to the variety of LCIA methods available in literature and the complexity in choosing one for an LCA study, there is a demand by LCA practitioners for support in decision making in private and public organizations. Therefore, the Brazilian Life Cycle Impact Assessment Network (Rede de Pesquisa em Avaliação do Ciclo de Vida, RAICV, 2014), (Regimento da Rede de Pesquisa em Avaliação de Impacto do Ciclo de Vida. São Bernardo do Campo, 11 November 2014, RAICV) evaluated different LCIA methods, for several impact categories, including Abiotic Resources. Nevertheless, due to certain characteristics from the Abiotic Resources category (e.g., a relative site-generic impact), the results for this category may be applied to other countries as well. As will be seen later, some RAM create characterization factors (CF) for both biotic and abiotic resources; thus, in some cases the recommendation went beyond abiotic resources (for RAM). Therefore, the objective of this manuscript was to evaluate different operational LCIA methods for (abiotic) resources available in literature in order to propose a recommendation .To facilitate the assessment, we applied some of these operational LCIA methods to a case study of ethylene production in Brazil through bio-based and fossil-based routes.

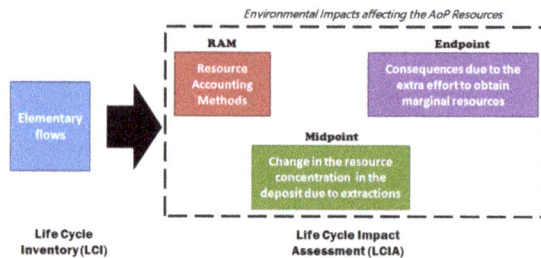

Figure 1. Simplified representation of cause-effect on AoP Resources, based on traditional LCIA groupings proposed by ILCD [3] and Swart *et al.* [2]

2. Results and Discussion

2.1. Operational Resource-Based LCIA Methods

For the evaluation of the LCIA methods, we adopted the classification according to ILCD [3] and Swart *et al.* [2], *i.e.*, in three levels of impact assessment (Figure 1): RAM, midpoint, and endpoint methods. In total, we found 19 operational LCIA methods, which are described and assessed below. The values for each LCIA method can be seen in Table 1 (for RAM), Table 2 (for midpoint), and Table 3 (for endpoint).

2.1.1. Resource Accounting Methods (RAM)

CED

The Cumulative Energy Demand (CED) [13] is one example of an operational LCIA method for quantifying the cumulative energy use, which was introduced in the 1970s by Boustead and Hancock [14] and Pimentel *et al.* [15], and standardized by VDI (the Association of German Engineers) [16]. It is an RAM that uses the heating value of materials as an aggregation unit. Frischknecht *et al.* [17] evaluated several operational LCIA methods with this approach, pointing out some differences among them, such as the use of high or low heating values. The CED can also be used as a proxy in LCA, since it presented a direct correlation with several other LCIA indicators [18,19], especially for the fossil energy category. Generally, CED (and similar methods) accounts solely for the resources that have a certain energy or heating value, which may be seen as a limitation of the approach because it becomes restricted to energetic resources (fossil, nuclear, solar, geothermal, wind, and hydropower) and biomass. It is not a regionalized method, but according to Alvarenga [20], spatial-differentiation may be considered for land use if an adaptation is previously performed to avoid double-counting with biomass, as suggested in Alvarenga *et al.* [21]. Since it uses energy, based on the first law of thermodynamics, the scientific robustness is not so high since there are other LCIA methods that use the second law of thermodynamics, which has a higher scientific robustness (see below). Further, since it has CF entirely for energy resources and biomass (and site-generic), the CED had low scores for both criteria.

CExD

The Cumulative Exergy Demand (CExD) [22] is a RAM that uses the CED as a baseline, but instead of using energy as the indicator, it uses exergy. Exergy of a resource or system is the maximum amount of useful work that can be obtained from it [23]. By using exergy, the CExD is able to account for several types of resources (fossil, nuclear, wind, solar, potential, biomass, water, metals, and minerals), including those with no heating value. Land use is not accounted for in order to avoid double-counting with biomass. This method does not have spatial-differentiated CF, but according to Alvarenga [20], it may be adapted to consider the regionalization for land use (as in CED). Dewulf *et al.* [9] and Swart *et al.* [2] mentioned several inconsistencies in the CExD model to calculate the exergy value of metals, minerals, and biotic resources. Since CExD uses exergy (2nd law of thermodynamics), it has a higher scientific robustness than CED and a higher amount of CF. Therefore, it had a medium score (higher than CED) for the criteria evaluated.

CEENE

The CEENE method [9] is a RAM that tries to aggregate different types of resources in a single unit (exergy). It considers several elementary flows, with a higher number of CF than other RAM that use the same approach (e.g., CExD). Moreover, it is seen as an evolution of CExD, and by using exergy instead of energy, it has a higher scientific robustness than CED. Liao *et al.* [8] considered CEENE as (one of) the best thermodynamic-based LCIA method(s). One of the differences between CEENE and CExD is the approach used to account for the exergy of metals and minerals and for biotic resources (biomass and/or land use). CEENE 1.0 does not have spatially-differentiated CF; however, in CEENE 2.0 (There is an even more recent version (3.0), where there is also spatial-differentiation for ocean occupation (e.g., for aquaculture), but it was not considered since it was published after 2014), there is spatial-differentiation for land use through the work of Alvarenga *et al.* [21]. In this case, CF are presented in different scales (see in below). Due to the advance in the model to calculate exergy for some resources, CEENE has a higher score for scientific robustness than CExD and higher number of CF. Further, CEENE 2.0 has regionalized CF for land use. Therefore, CEENE had the highest score for both criteria, in comparison to other RAM.

SED

The Solar Energy Demand (SED), developed by Rugani *et al.* [10], is a RAM that uses the Emergy concept [24] as a baseline. In Emergy, the cradle of an LCA is not in the boundary between the ecosphere and anthroposphere, as considered by several RAM (e.g., CEENE), but within the limits of the geobiosphere, *i.e.*, the Sun, tidal energy, and geothermal energy, aggregated into an indicator called solar energy equivalent [2]. In this LCIA method, the authors focused on creating a high amount of CF through the compilation of several published works that had quantified the transformities (Transformity is the name give in Emergy scientific community to what is called as CF in LCA community) of different natural resources. Regarding the possible double-counting between biotic resources and land use, SED follows the approach of CEENE, which chooses to account for land use (in contrary to CED and CExD); therefore, we can say that SED and CEENE are equivalent in this aspect. As previously mentioned, SED has a high number of CF, but they are not regionalized. Emergy is an approach that is still challenged by the scientific community, even though there are some efforts to solve some of the problems and align it to LCA [25,26], in which SED was a result of that effort. Due to those reasons, SED received a medium-high score for both criteria.

MIPS

Material Inputs Per Service (MIPS) [27,28] is an indicator of the cumulative amount of resources of product/service through its life cycle, and sometimes is called a Material Footprint. MIPS is based on material flow analysis, separating materials in five classes: abiotic resources, biotic resources, earth movement, water, and air [28]. As some other RAM, MIPS was developed outside of LCA and later it was considered as an LCIA method. Saurat and Ritthoff [29] proposed a method to calculate CF for ecoinvent database v2.2. However, the CF are not fully made available in that publication, while the authors mention that the CF are a beta version and the new CF shall be available in Wuppertal Institute website in the future. Even though in Ritthouff *et al.* [28] and in Saurat and Ritthoff [29], the possibility and implementation of regionalization is mentioned, this is in fact a regionalization of LCI, and not LCIA as in CEENE 2.0. Therefore, MIPS does not have regionalized CF. Saurat and Ritthoff [29] and Wiesen *et al.* [30] mentioned some differences between traditional MIPS assessment and LCA and the need for adaptation of MIPS into an ecoinvent database, as the lack of elementary flows from unused extracted materials from mining. Due to those reasons, MIPS received a medium-low score for scientific robustness and a medium score for scope.

LREx

The Exergy-based accounting for land resource (LREx) [21] is a method focused on accounting for land use as a natural resource through exergy, and is proposed to be complementary to the CEENE method. Regarding the issue of avoiding double-counting in biotic resources in RAM, LREx suggests to account for the biomass extracted, when this originally occurs in a natural environment, and for the land use, when originally changed in a man-made environment, and in this case, based on natural potential net primary production (NPP). Regarding the latter, it has spatial-differentiated CF in different scales, *i.e.*, site-generic (world), continent-based, country-based, region-based (for six countries), and on a grid scale (approximately 10 km × 10 km). In this sense, LREx had a high scientific robustness, but a low score for scope, even though it was regionalized, since it was focused solely on land use and biotic resources. In fact, LREx is a specific LCIA method for a type of elementary flow (land use); thus, it is proposed to be used as complementary to other RAM, and not as a single indicator.

ICEC/ECEC

The Industrial Cumulative Exergy Consumption/Ecological Cumulative Exergy Consumption (ICEC/ECEC) is an LCIA method developed by Hau e Bakshi [31] and Zhang *et al.* [32], based on exergy. The method has CF that are operational for extended input-output databases, as USA Input-Output Database 1997, while other LCIA methods (previously mentioned) are operational for process-based LCI (e.g., ecoinvent). First, the authors tried to operationalize the cumulative exergy consumption (CExC), proposed by Szargut [33], via the ICEC. However, this method also proposes an additional approach, filling the gap between LCA and economic assessment of natural resources (evaluating ecosystem services), via the ECEC [32]. For the latter, it used principles from Emergy [24], allowing for consideration of the exergy consumption of ecological goods and services in solar energy equivalents. Therefore, ECEC has a similar approach to SED, where the cradle is at the boundary of the geobiosphere. Even though ICEC/ECEC tries to include ecosystem services, this inclusion is still limited for some important ones, as pollination and carbon sequestration. Moreover, ICEC/ECEC does not have regionalized CF. Nevertheless, ECEC proposes some solutions for emergy critical aspects, such as allocation. For those reasons, ICEC/ECEC had a medium-high score for both criteria.

EF

The Ecological Footprint (EF) is defined as the area of water and land needed to directly and indirectly support a certain population [34]. It divides these areas in six classes: crop land, forest, pasture, water, infrastructure, and energy (carbon sequestration). For LCA, EF of a product may be defined by the area directly and indirectly needed during the life cycle of this product. EF was operationalized as an LCIA method by Huijbregts *et al.* [35], where it was also added (through some adaptations) to the area needed for nuclear resources. EF appeared to be a good proxy, but it was not recommended for mineral-based products or those with high particulate matter emission [35]. It may be interpreted as complementary to CED/CExD, making it possible to account for land use with low uncertainty, but for that, some adaptations are needed to avoid double counting with biotic resources. EF may be regionalized for different scales, based on specific biocapacities, but there is no operational LCIA method with spatially-differentiated CF. On the other hand, EF (as an LCIA method) only considers elementary flows related to fossil and nuclear resources, land use, and CO_2 emission. Therefore, it has a limited scope (it does not account for metals and minerals), and goes beyond the AoP Resources by accounting for CO_2 emissions. For these reasons, the final score of this RAM was rather low.

Table 1. Quali-quantitative assessment of the LCIA methods at the RAM level for the AoP Resources.

LCIA Method	CEENE	CExD	CED	SED	MIPS	LREx	ICEC/ECEC	EF
Base reference	[9]	[22]	[13,16]	[10]	[28]	[21]	[31,32]	[34,35]
Criterion #1 (Scope)	5	3	1	4	3	1	4	3
Criterion #2 (Scientific robustness)	5	3	2	4	2	5	4	2
Final score	5.0	3.0	1.5	4.0	2.5	3.0	4.0	2.5
Observation	We evaluated v1.0 and v2.0	-	-		-	Specific for land use	-	-

2.1.2. Midpoint LCIA Methods

ADP

Abiotic Depletion Potential (ADP) was initially developed by Guinée [11], was later modified in van Oers *et al.* [36], and included in the LCIA methodology CML-IA (from the Institute of Environmental Sciences (CML), named CML-IA). To calculate CF, ADP uses an equation that involves the extraction rate of a certain resource and the squared of the availability of this resource in deposits. All resources are normalized to the CF of antimony (Sb); therefore, the indicator is Sb equivalent (Sb-eq). Regarding the deposits used to calculate the CF, original ADP used the ultimate reserves, which may be defined as the total amount of a certain substance (e.g., iron) available in the Earth's crust, oceans and atmosphere. In this sense, the ultimate reserve also includes deposits that are not economically or technically feasible for extraction. Later, ADP created CF for other approaches as well, *i.e.*, for reserve base and economic reserves. ADP has CF for metals and minerals, and the total amount of CF is dependent on the approach used (ultimate reserve, reserve base or economic reserve). For fossil fuels, ADP also has CF, but in the latter versions of the ultimate reserve approach (e.g., v4.2), the CF are based on the net heating value of the fossil fuel, similar to CED (a RAM), thus not providing a midpoint resource depletion assessment. ADP has high scientific robustness in LCA community, and the approach based on the reserve base is recommended by ILCD [3] as the LCIA method to be used for midpoint assessment. Furthermore, the amount of CF provided by this method is quite high in comparison to other depletion LCIA methods, despite the approach considered. In this sense, ADP received high scores for the criteria considered.

EDIP

EDIP (Environmental Design of Industrial Products) 1997/2003 [37] is an LCIA methodology established in the LCA community and traditionally used for the assessment of products. This LCIA methodology considers several impact categories, including the depletion of resources, with CF for metals, minerals, and fossil fuels. For this category, the CF are calculated by an equation that exclusively involves the amount of available resources in deposits, *i.e.*, not considering the extraction rate, as in ADP. Therefore, the property of a resource for having a high/low extraction rate is not accounted for by the CF. For this reason, EDIP received a lower score than ADP in the criterion scientific robustness. The deposits considered in this LCIA method are based on the economic reserves. EDIP has a high amount of CF for the resources previously mentioned, higher than Recipe Midpoint [38], but slightly lower than ADP (depending on the approach used). For that reason, EDIP had a medium-high score for scope.

Recipe Midpoint

Recipe Midpoint (Recipe has two versions (midpoint and endpoint) and, regarding resource depletion, there are differences in the analysis. For this reason they were evaluated separately.) [39] has CF for fossil fuels, metals and minerals, and the approach used is different for the types of resources. For fossil fuels, Recipe Midpoint considers the heating value, *i.e.*, a RAM approach, similar to CED. For metals and minerals, the approach is different, considering the depletion of those resources at midpoint level. Recipe Midpoint has an innovative approach, in comparison to ADP and EDIP, evaluated through the deposits of minerals, and not by the metals *per se*. According to the authors, by doing this, the LCIA method better represents the reality of the metals' geological distribution, allowing them to cover a higher number of commodities, especially those extracted as by-products. This method considers the change in the ore grade, *i.e.*, the decrease in the concentration of minerals in an ore due to extraction. Recipe Midpoint has less CF than ADP and EDIP, and this is probably due to the higher complexity level to obtain the data needed to create the CF by the approach from the former. Further, Recipe Midpoint has some inconsistencies in the creation of CF [40] (e.g., allocation procedure). For those reasons, this method had a medium score for both criteria.

ORI

ORI (Ore Requirement Indicator) [40] is a specific LCIA method for the evaluation of metal and mineral depletion; therefore, it does not have CF for fossil fuels, for instance. It uses a similar approach to Recipe Midpoint, *i.e.*, considers the change in ore grade, and its equation is the reciprocal of the ore grade variation, thus the change in ore mass by mass of metal extracted. However, instead of using a small database (locally or temporally) to generate CF, as performed in Recipe Midpoint, the authors use a more robust database that has information from different mines for more than a decade. In order to guarantee scientific soundness, the authors created CF solely for the metals in which the source from the database represented more than 50% of worldwide production. In this way, it has a higher scientific robustness than Recipe Midpoint. On the other hand, it had a limited number of CF (9 metals). ORI has an interesting approach, indicating an option to be followed for midpoint assessment, especially with the expansion of data for metals and minerals. However, since it has solely 9 CF and is focused on metals and minerals (no CF for fossil fuels), it has a low operationalization for LCA (later this LCIA method may be enhanced by the inclusion of more CF), and as a consequence, it had a low score for scope and a high score for scientific robustness.

AADP

The Anthropogenic Stock Extended Abiotic Depletion Potential (AADP) [41] may be considered a complementary method for the ADP, by including the depletion assessment resources that have already been extracted from their deposits and are now available in the anthroposphere (e.g., landfill), bringing an innovative concept. However, due to the difficulty of obtaining consistent data, it has CF for solely 10 metals. Since it is a specific LCIA method for metals, it does not have CF for fossil fuels. For those reasons, it had a high score for scientific robustness and a low score for scope, as ORI. AADP went through some adaptations, where more CF were created (35 in total), but this new version [42] was not considered in this study because it was published after December 2014.

OGD

The Ore Grade Decrease (OGD) was proposed by Vieira *et al.* [43], and it has a similar approach to ORI, *i.e.*, it evaluated the ore grade change due to extraction of metals, based on a geologic distribution model. For this reason, it has a high scientific robustness, but the authors created only one CF, for copper, giving a low score for the scope criterion. In order to differentiate this from other low-scoring LCIA methods at the scope criterion (ORI and AADP), we gave an even lower score for OGD.

Table 2. Quali-quantitative assessment of the LCIA methods at midpoint level for the AoP Resources.

LCIA Method	ADP	EDIP	Recipe Midpoint	ORI	AADP	OGD
Base reference	[11,36]	[37]	[39]	[40]	[41]	[43]
Criterion #1 (Scope)	5	4	3	1	1	0.5
Criterion #2 (Scientific robustness)	4	3	3	5	5	5
Final score	4.5	3.5	3.0	3.0	3.0	2.75
Observation	-	-	-	Specific for metals	Specific for metals at anthroposphere	Since it had only one CF

2.1.3. Endpoint LCIA Methods

Eco-Indicator 99

Eco-Indicator 99 is an LCIA endpoint method, *i.e.*, it evaluates the final impact from using fossil fuels, metals and minerals by the approach proposed in Muller-Wenk [4], which considers the increased (future) workload in the extraction of more inaccessible reserves resources (e.g., marginal reserves). In Eco-indicator 99, the surplus energy (MJ_{se}) is used as an aggregated unit for resource depletion [44]. This resource depletion characterization model is used in other LCIA methodologies as well, such as in TRACI (the Tool for the Reduction and Assessment of Chemical and other environmental Impacts), BEES (Building for Environmental and Economic Sustainability), and Impact 2002+ [13]. The base for the Eco-indicator 99 assessment is through the estimated surplus energy required for the extraction of minerals, based on the decline of the concentration rate in time [4,45]. The analysis uses geostatistical models to indicate the distribution structure, quantity and quality of the minerals and the future effort needed, to calculate the surplus energy for extraction of resources. Following the same trend from other endpoint LCIA methods, the Eco-indicator 99 has fewer CF than midpoint LCIA methods. In addition, compared to other endpoint LCIA methods (e.g., Recipe Endpoint and EPS2000), the number of CF is also lower, and with the absence of important minerals in the global context, such as gold, iron, palladium, molybdenum, lead and platinum. For this reason, the score for scope was medium. On one hand, the characterization model is part of different methods LCIA (e.g., Impact 2002+), which are consolidated and accepted by the scientific community, but on the other hand, the characterization models was based (for metals) on a limited number of low accuracy curves. Therefore, Eco-indicator 99's scientific robustness was considered medium-low, in comparison to other endpoint LCIA methods.

Recipe Endpoint

Recipe Endpoint [39] assesses the depletion of fossil fuels, metals and minerals through another approach (compared to its midpoint version), going beyond the cause-effect relationship, *i.e.*, to an endpoint level [2,3]. The approach, somewhat similar to the Eco-indicator 99, evaluates the increase in the cost of extracting those resources due to their depletion, thus using the approach proposed by Müller-Wenk [4] and Stewart and Weidema [5]. Because it is an LCIA method that goes beyond the cause and effect of resource depletion impacts, it has greater complexity of data to create CF, and in this sense, it has fewer CF than midpoint LCIA methods (e.g., ADP). Moreover, when compared to other endpoint LCIA methods, it has more CF than the Eco-indicator 99, for instance. For metals and minerals, CF were calculated based on extraction costs and on the CF of Recipe Midpoint. Therefore, the endpoint CF also has the same calculation inconsistencies mentioned in Swart and Dewulf [40] (e.g., allocation procedure). Regarding fossil fuels, the CF for crude oil was based on data from the International Energy Agency, assuming constant annual production over time, using a limited relationship of price and production and determining an arbitrary time period [46]. Additionally, due to a lack of data, coal and natural gas CF were calculated using extrapolated data from crude oil. Thus, the Recipe Endpoint received a medium-high score for the scope criterion and a medium score for the scientific robustness criterion.

EPS2000

The Environmental Priority Strategies (EPS) is an endpoint LCIA method, proposed for the first time in 1990, and it was later modified; the final version is named EPS2000 [47,48]. The principle for damage assessment is through the willingness-to-pay (WTP) approach, in which natural resources and environmental impacts are put in monetary values. One of the impact categories is named abiotic stock resource (or depletion of reserves), which evaluates the depletion impacts (at endpoint level) from metals, minerals and fossil resources. The unit used is the Environmental Load Unit (ELU), which represents the costs of sustainable extraction of non-renewable resources. Considering the limitations of WTP for abiotic resources, EPS2000 proposes a market scenario, in which the production costs of a certain substance are used to estimate the CF for resource depletion. The characterization models from EPS are not entirely transparent, considering that they are based on political and sociocultural values. Therefore, EPS2000 had a medium-low score for scientific robustness. The current version of EPS2000 has a significant amount of CF, higher than other endpoint LCIA methods (e.g., Recipe Endpoint), and for this reason it received a high score for scope. As a consequence, EPS2000 received the same final score from Recipe Endpoint.

SuCo

The Surplus Cost (SuCo) method, proposed in Ponsioen *et al.* [46], aims to adapt or integrate elementary flows related to fossil resources to LCA. Therefore, this LCIA method is specific for that type of resource, not generating CF for metals and minerals (amongst other resource types). The proposal from SuCo is to assess fossil resource scarcity based on the future increase in global costs due to the use of marginal fossil resources used in the life cycle of products. Therefore, SuCo follows the same trend as Eco-indicator 99 and Recipe Endpoint. In practice, SuCo may be seen as an evolution from Recipe Endpoint (which may be seen as an evolution from Eco-indicator 99), and that there is a possibility to incorporate SuCo to Recipe Endpoint in the next versions of the latter [49]. SuCo has CF for three types of fossil fuels (crude oil, natural gas, and coal), from which it used specific data for the calculation of the respective resource. Therefore, it has higher scientific robustness than Recipe Endpoint, in which only data from crude oil was used. For that reason, it received a medium-high score for that criterion. Since it is a specific LCIA method for fossil fuels, it received a low score for scope.

Exergoecology

Exergoecology, proposed by Valero e Valero [50,51], brings an innovative approach to the LCA community, where the authors try to quantify the depletion of metals and minerals through the exergy cost, which, in a simplified way, would be to make a cumulative exergy consumption assessment, but in the opposite direction, *i.e.*, from grave-to-cradle. The main idea is that the method quantifies the exergy needed to let the metal be ready for extraction through mining, from its reference state (where exergy is zero). This method may be considered as evaluating the AoP Resources at the endpoint level, since it considers the consequences from resource depletion. Exergoecology has an interesting alternative approach, but it is not yet completely operational for LCA, mainly for two reasons: (1) Through the published articles, we were able to quantify only seven CF; and (2) this originates from a scientific area that needs more research, and more CF should be generated. Regarding the latter reason, even though exergy is already established in LCA, the scientific proposal from Valero and Valero [50,51] goes beyond traditional exergy, *i.e.*, they propose accounting for the exergy from grave-to-cradle. For those reasons, Exergoecology received a low score for scope and a medium-high score for scientific robustness.

Table 3. Quali-quantitative assessment of the LCIA methods at the endpoint level for the AoP Resources.

LCIA Method	Eco-Indicator 99	Recipe Endpoint	EPS2000	SuCo	Exergoecology
Base reference	[44]	[39]	[47,48]	[46]	[50,51]
Criterion #1 (Scope)	3	4	5	1	1
Criterion #2 (Scientific robustness)	2	3	2	4	4
Final score	2.5	3.5	3.5	2.5	2.5

2.2. Case Study

We applied an ethylene production case study to some of the aforementioned LCIA methods. We compared the traditional fossil-based ethylene (FE) to the bioethanol-based ethylene (BE), produced from sugarcane. From the 19 different LCIA methods, 13 were selected to be applied in the case study. Six LCIA methods were excluded due to the lack of operational CF. The LREx method was applied via the CEENE v2.0 method. Further, it is interesting to note that some LCIA methodologies (Impact 2002+, BEES, and TRACI) are indirectly considered in this study since they use a similar characterization model as Eco-indicator 99. It may be important to mention that the LCIA method Ecological Scarcity [52] was not considered in the case study, nor in the previous theoretical assessment, because it is a Swiss-based distance-to-the-target method, and is not in the scope of the Brazilian context. For ADP, we used three versions in the assessment of better evaluation: (1) ADP v3.2, an older version of ADP that accounts for metals, minerals, and fossil fuels in Sb-eq, through the ultimate reserves (thus midpoint assessment); (2) ADP v4.2, a newer version of ADP that accounts for metals and minerals as Sb-eq, through ultimate reserves, while for fossil fuels it accounts for the low heating value (thus similar to a RAM); and (3) ADP-ILCD, an ADP version that assesses metals, minerals and fossil fuels as Sb-eq through a reserve base.

2.2.1. RAM

For all five RAM methods, the results showed BE with higher environmental impacts than FE, as can be seen in Figure 2. For the CED, CExD, CEENE v2.0, and SED, FE had approximately half of the total value of BE (varying between 47% and 54%). On the other hand, FE had approximately 34% of the total value of BE for the CEENE method v1.0. The main hotspots found in each of the RAM methods are discussed below.

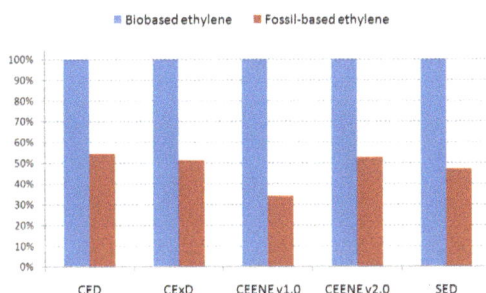

Figure 2. Resource-based assessment of bio-based ethylene and fossil-based ethylene, with different RAM methods, normalized to the highest value

CED and CExD had similar results, *i.e.*, BE had greater environmental impact, mainly due to the energy content of the sugarcane (86%–87%), and a considerable fraction of the total value was due to fossil energy consumption (10%–11%), mainly due to natural gas consumed during ethanol-to-ethylene production and diesel consumed in the sugarcane stage. For the FE, most of the environmental impacts were due to the crude oil (63%–64%) and natural gas (31%–33%) consumption in the ethylene supply chain. An interesting difference between CED and CExD was the results for water resources, *i.e.*, for CED, water resources were consumed mainly due to hydropower (potential energy), while the CExD accounted not only for water from hydropower (potential energy), but water was also used in the ethanol production process.

For the CEENE method, BE environmental impacts were mainly due to land occupation for the sugarcane cultivation (90% in v1.0 and 85% in v2.0). Fossil fuels also had a significant contribution in the total value (7% in v1.0 and 11% in v2.0), and from that approximately 53% was due to natural gas consumption in the ethanol-to-ethylene production process and 22% was due to diesel consumption during sugarcane cultivation. The main difference between v1.0 and v2.0 was land occupation, *i.e.*, while the former uses average European solar irradiation as a proxy, giving a higher result, the latter uses regionalized data on natural potential NPP as a proxy (which for this case study, gave a lower value).

When using the SED, BE had more than double the total environmental impacts, and the main hotspot was the consumption of gypsum (mineral) in the sugarcane stage, which was responsible for more than 66% of the total environmental impact. After that, the consumption of limestone (mineral), with 10% of total, natural gas in the ethanol-to-ethylene process (4%), and diesel at the sugarcane stage (3%) were also relevant contributors for the SED. Regarding the FE, most of the environmental impacts were due to crude oil (67%) and natural gas (26%) consumed in the ethylene supply chain. Sodium chloride had also a relevant share of contribution (3%) for FE.

It is interesting to note that for FE, the results among the five RAM showed similar hotspots. On the other hand, while for CED, CExD, CEENE v1.0 and CEENE v2.0, the main hotspot for BE was the sugarcane (either as a biomass or the land occupation for its cultivation); in SED the land occupation impacts accounted for only 2% of the total environmental impacts, and the minerals consumed during the sugarcane stage were the main hotspots. This is due to the different approach used in SED, which sets the geobiosphere as system boundaries to create the CF, while the former four RAM set the boundary between the ecosphere and the anthroposphere [2,21].

2.2.2. Midpoint

For the midpoint level, the results were slightly divergent among the LCIA methods (Figure 3). For ADP v3.2, ADP v4.2, and the Recipe Midpoint, BE was the most beneficial option for the environment, with values of approximately 20%–30% of the environmental impact of FE. On the other hand, ADP-ILCD, EDIP2003, ORI and AADP had opposite results, and the degree of how beneficial FE was varied considerably. The main reason was the higher importance given to metals and minerals in the latter LCIA methods. The relevance of the elementary flows for each LCIA method can be better visualized in Table 4, Figures 4 and 5.

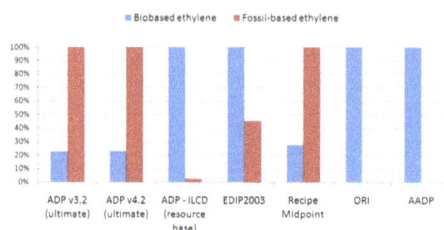

Figure 3. Resource-based assessment of bio-based ethylene and fossil-based ethylene, with different midpoint LCIA methods, normalized to the highest value.

Table 4. Relative contribution of specific elementary flows for bio-based ethylene (BE) and fossil-based ethylene (FE) at the midpoint and endpoint LCIA methods considered in this case study.

Elementary Flows (Natural Resources)	ADP v3.2 (Ultimate)		ADP v4.2 (Ultimate)		ADP-ILCD (Res. Base)		EDIP2003		Recipe Midpoint		ORI		AADP		Eco-Indicator 99		Recipe Endpoint		EPS2000	
	BE	FE	BE	FE	BE	FE	BE	FE	BE	FE	BE	FE	BE	FE	BE	FE	BE	FE	BE	FE
Metals and minerals																				
Chromium	-	-	-	-	-	-	-	-	2%	-	-	-	-	-	-	-	1%	-	2%	-
Copper	-	-	2%	-	4%	-	7%	-	4%	-	34%	28%	-	-	1%	-	2%	-	13%	-
Iron	-	-	-	-	-	-	-	-	6%	-	-	-	-	-	-	-	2%	-	2%	-
Lead	-	-	1%	-	30%	-	12%	-	-	-	1%	1%	-	1%	1%	-	-	-	7%	-
Nickel	-	-	-	-	4%	-	35%	-	3%	-	57%	61%	-	1%	1%	-	1%	-	14%	-
Uranium	-	-	-	-	-	49%	-	8%	-	1%	-	-	89%	92%	-	-	-	-	-	-
Phosphate rock	-	-	-	-	1%	-	-	-	-	-	-	-	-	-	-	-	-	-	3%	-
Zinc	-	-	2%	-	58%	-	23%	-	-	-	1%	-	2%	-	-	-	-	-	14%	-
Fossil fuels																				
Crude oil (diesel-cane)	20%	-	21%	-	-	-	2%	-	14%	-	-	-	-	-	23%	-	22%	-	5%	-
Crude oil (FE)	-	36%	-	34%	-	28%	-	59%	-	66%	-	-	-	-	-	66%	-	66%	-	51%
Crude oil (other)	8%	-	8%	-	-	-	1%	-	11%	-	-	-	-	-	9%	-	8%	-	3%	-
Natural gas (BE)	54%	-	51%	-	-	-	5%	-	40%	-	-	-	-	-	55%	-	50%	-	25%	-
Natural gas (FE)	-	62%	-	65%	-	15%	-	32%	-	32%	-	-	-	-	-	34%	-	32%	-	49%
Natural gas (other)	8%	-	7%	-	1%	-	1%	-	6%	-	-	-	-	-	8%	-	7%	-	4%	-
Hard coal	6%	-	4%	-	-	-	-	-	4%	-	-	-	-	-	-	-	4%	-	-	-
Other elementary flows	4%	2%	6%	1%	3%	8%	14%	1%	10%	1%	7%	10%	9%	6%	4%	0%	5%	2%	8%	0%

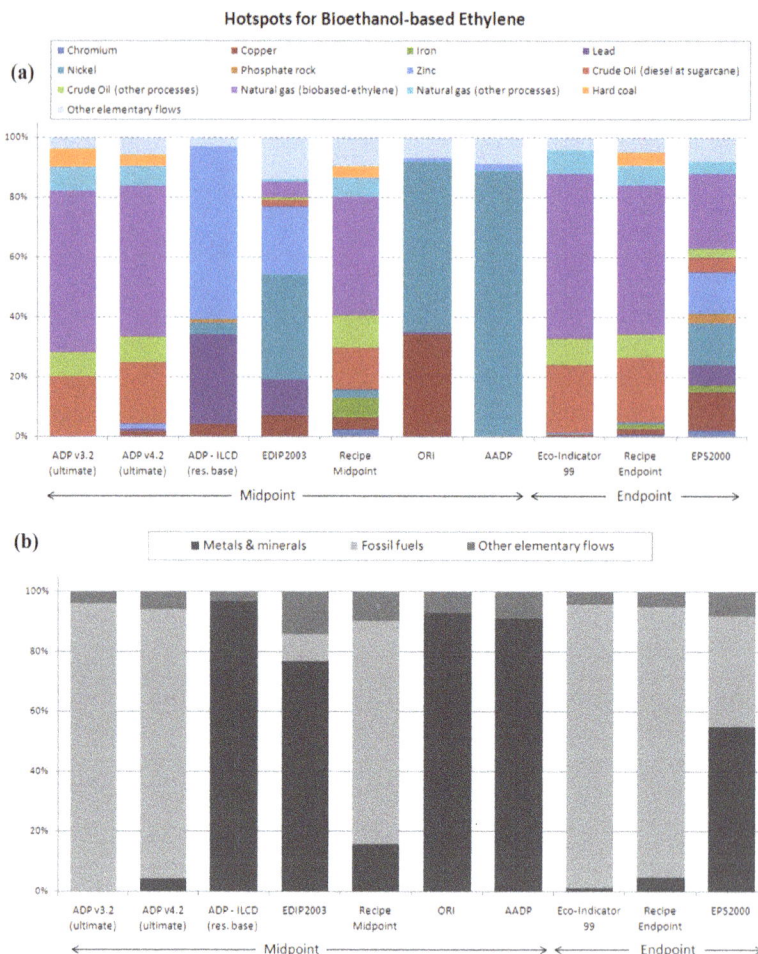

Figure 4. Hotspots from bioethanol-based ethylene with respect to (**a**) elementary flow level; and (**b**) type of resource level for midpoint and endpoint LCIA methods.

For the two versions of ADP that consider the ultimate reserve as background data for CF (v3.2 and v4.2), fossil fuels seemed to be much more relevant than metals and minerals. As a consequence, FE had worse results, and the main driver for that was the consumption of crude oil (62% for ADP v3.2 and 65% for ADP v4.2) and natural gas (36% for ADP v3.2 and 34% for ADP v4.2) at the ethylene supply chain. Meanwhile, for BE, the main hotspots were also fossil fuels, *i.e.*, natural gas consumed at the ethanol-to-ethylene process (54% of the environmental impacts in both methods) and crude oil consumed as diesel at the sugarcane stage. On the other hand, ADP-ILCD, which considers the reserve base approach, seemed to give more importance to metals and minerals, putting BE as the product with the highest environmental impacts, from which zinc, lead, copper, and nickel, mainly used in the production of agricultural machinery, were the main hotspots for BE. Meanwhile, due to the approach considered (reserve base), fossil fuels used in the agricultural stage had a minor contribution. For FE, uranium (This is a limitation of the study; the ecoinvent dataset was used for FE, which is European-based, thus has Uranium is an important energy source for electricity), natural gas and crude oil were the main hotspots, all of which were consumed in the ethylene supply chain (Table 4, Figures 4 and 5).

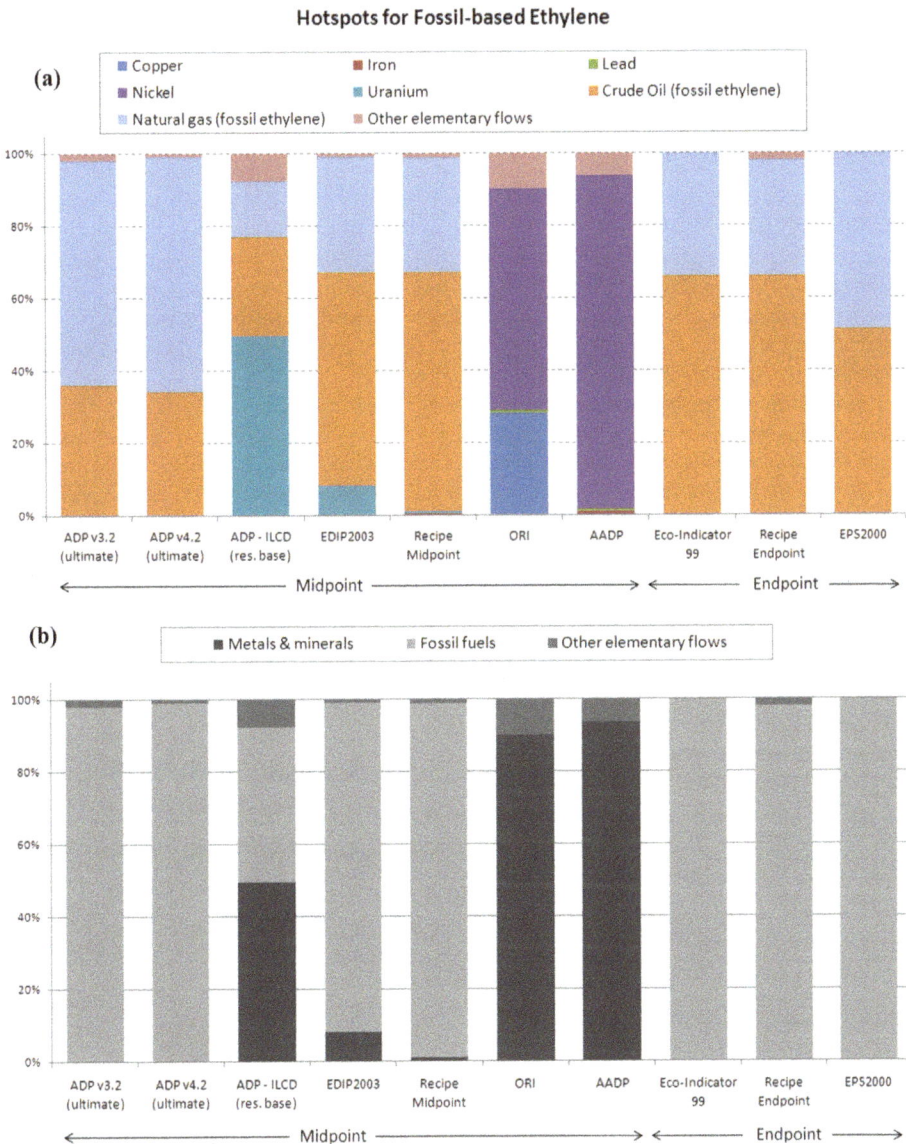

Figure 5. Hotspots from fossil-based ethylene with respect to (**a**) elementary flow level; and (**b**) type of resource level for midpoint and endpoint LCIA methods.

Similar to ADP v3.2 and v4.2, the Recipe Midpoint also gave more importance to fossil fuels (after normalization), identifying FE as the product with the highest environmental impacts. Crude oil (66%) and natural gas (32%) consumed in the ethylene supply chain were the main hotspots for FE. The main hotspots for BE were natural gas (46%) and crude oil (25%), consumed in several processes. Moreover, some metals also had a considerable contribution to BE, especially chromium, copper, iron, and nickel (summing to 15% of the environmental impacts), mainly used in the production of agricultural and industrial machinery (Table 4, Figures 4 and 5).

EDIP 2003 seemed to give more importance to metals and minerals, probably due to the economic reserve approach. BE was the product with the highest environmental impacts, mainly due to nickel, zinc, lead, and copper (total of 77%), primarily used in the production of agricultural and industrial machinery. Similar to ADP-ILCD, fossil fuels used in the agricultural stage had minor contributions. The main hotspots for FE were crude oil (59%) and natural gas (32%) consumed in the ethylene supply chain (Table 4, Figures 4 and 5).

Regarding ORI and AADP, BE was the product with the highest environmental impacts, since these methods have CF solely for metals and minerals thus far. In AADP, BE was highly influenced by nickel (89%), while in ORI it was mainly influenced by nickel (57%) and copper (34%). The same metals were the main hotspots for FE, showing that a low number of CF may indicate a misleading interpretation (if we consider the results from the other midpoint LCIA methods). In this sense, it is important to mention that these LCIA methods may be currently used in LCA with caution, *i.e.*, it should be complemented by other LCIA methods for other types of resources (e.g., fossil fuels), and avoid making a joint overall abiotic resource assessment.

From the results at the midpoint level, we could see that certain LCIA methods gave more focus to metals and minerals, as those considering a reserve base (ADP-ILCD) and an economic reserve (EDIP2003) as the background approach for creating CF. This is probably due to the lower amount of metal deposits available in higher concentrations. Meanwhile, ADP versions considering the ultimate reserves (v3.2 and v4.2) and the Recipe Midpoint gave more focus to fossil fuels. Furthermore, we could notice that using LCIA methods with a low amount of CF, such as ORI and AADP, may show inconsistent results, especially for products highly based on fossil fuels (as FE). Therefore, this case study highlights the importance of the criterion assessing the number of reference flows with CF, when dealing with abiotic resource depletion. In this sense, it is important to highlight that a possible procedure to choose an LCIA method is to make a preliminary environmental impact assessment, finding possible hotspots (e.g., iron may be a hotspot when performing an LCA of industrial machinery), and then making sure that the chosen LCIA method has CF for those elementary flows.

2.2.3. Endpoint

For the endpoint LCIA methods, the results of the case study were more convergent than the midpoint LCIA methods, *i.e.*, they all considered the BE as more beneficial to the environment (Figure 6).

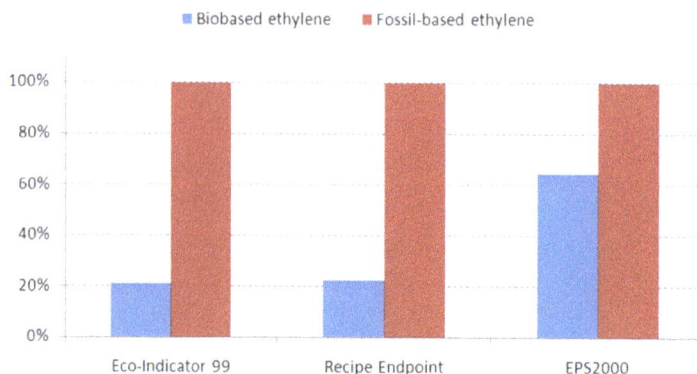

Figure 6. Resource-based assessment of bio-based ethylene and fossil-based ethylene, with different endpoint LCIA methods, normalized to the highest value.

For the Eco-indicator 99, the environmental impacts from FE were mainly due to crude oil (66%) and natural gas (34%) consumed in the ethylene supply chain. Thus, the environmental impacts from metals and minerals were negligible. Meanwhile, for the BE, 63% of the environmental impacts were due to overall natural gas consumption (87% of those in the ethanol-to-ethylene process), 32% due to crude oil consumption (72% of those due to diesel in the sugarcane stage), 1% from copper and 1% from nickel, both mainly due to agricultural machinery (Table 4, Figures 4 and 5).

The Recipe Endpoint had similar results as the Eco-indicator 99, *i.e.*, crude oil (66%) and natural gas (32%) consumed in the ethylene supply chain were the main hotspots for FE. For BE, the main hotspots were natural gas and crude oil, but some metals (chromium, copper, iron, and nickel) also had a significant contribution to the environmental impacts (Table 4, Figures 4 and 5).

The method EPS2000, which uses a different approach from the Recipe Endpoint and Eco-indicator 99, indicated different hotspots for BE. Natural gas and crude oil in different processes contributed to approximately 37% of the impacts, while several metals (chromium, copper, iron, lead, nickel, phosphate rock, and zinc) contributed to approximately 55% of the impacts (Table 4, Figures 4 and 5). For FE, the results were more similar to the other endpoint LCIA methods, *i.e.*, crude oil and natural gas consumed in the ethylene supply chain were the main hotspots.

2.2.4. Discussion

Liao *et al.* [8] analyzed different LCIA methods in a case study of titania produced in China through different routes (chloride and sulphate). Even though the products analyzed were quite different from our study, some results were similar among the RAM methods, showing that CED, CExD and CEENE were mainly influenced by fossil resources (since titania is not a bio-based product), while SED was mainly influenced by metals and minerals. Moreover, for the endpoint LCIA methods considered in Liao *et al.* [8], the results also highlight a higher contribution of fossil fuels for the Eco-indicator 99, in comparison to the EPS2000, which also had a significant contribution of metals. Therefore, Liao *et al.* [8] can corroborate our findings in both approaches (RAM and Endpoint).

Robech *et al.* [6] compared different resource-based LCIA methods by applying them to 2744 market datasets from the ecoinvent database. The results showed that the ADP (ultimate reserves), the Eco-indicator 99, and the Recipe Endpoint were mainly influenced by fossil fuels, while EDIP2003, ADP-ILCD and EPS2000 were mainly influenced by metals and minerals, corroborating the results of this study; except for EPS2000, which may have to do with the products analyzed, *i.e.*, our product system may be more fossil influenced than the averaged 2744 product systems analyzed by Robech *et al.* [6]. Further, even though our case study was still mainly influenced by fossil fuels in the EPS2000, the share of contribution was lower than the Recipe Endpoint and the Eco-indicator 99. For the RAM methods of Robech *et al.* [6], SED showed many more contributions to metals and minerals than CEENE v1.0 and CExD, validating the results of this study at the RAM level as well.

Moreover, similar results were found in the case study between the ADP (v3.2 and v4.2) and the Eco-indicator 99, identifying BE as the better option with approximately 20%–22% of the impacts from FE; this can be verified by the high correlation between these two LCIA methods found in Berger *et al.* [53]. Meanwhile, the similar results from our case study between ADP-ILCD and EDIP2003 (BE as the worst option), and among Eco-indicator 99, ADP (v4.2 and v3.2), and the Recipe Endpoint (BE as the best option) can also be corroborated by the high correlation found by Robech *et al.* [6] among those LCIA methods.

Klinglmair *et al.* [38] compared the CF of different resource-based LCIA methods, normalized with respect to iron, showing that ADP (It was not clear which version they used, but it seemed to be a version considering the ultimate reserve approach) has a higher relative CF for crude oil than other metals, confirming the results of this case study, and is mainly influenced by fossil fuels in ADP v3.2 and ADP v4.2. Meanwhile, the relative CF of crude oil (relative to iron) was not much higher for EPS2000 and the Recipe Endpoint, supporting our findings that showed iron with a significant contribution for those LCIA methods (Table 4, Figures 4 and 5), while the contribution of that metal in

ADP (v3.2 and v4.2) was negligible. On the other hand, according to Klinglmair *et al.* [38], Eco-indicator 99 has a lower relative CF for crude oil, similar to the Recipe Endpoint and EPS2000, but in the case study presented here, iron did not show significant contributions, and this is probably due to the specificities of the product system. The highest relative CF in AADP was from nickel [38], which confirms the reason why this metal was the main hotspot from our case study.

2.3. Recommendation of Abiotic Resource LCIA Methods

For each of one of the resource assessment levels (RAM, midpoint, endpoint), we proposed recommendations for the use of LCIA methods, based on the theoretical criteria assessment (Section 2.1), and with support from the case study (Section 2.2). As we can see in Table 1, CEENE v2.0 (CEENE with regionalized CF based on LREx) is the most suitable RAM due to the highest final score. However, we noticed that ICEC/ECEC also had a high final score; thus, since this method is operational for extended input-output LCI, we recommend it for that LCI approach; while CEENE v2.0 is recommended for process-based and hybrid LCI approaches. Further, based on Liao *et al.* [8], who recommends the SED method, and the concept introduced by ICEC/ECEC, which proposed complementary assessment for RAM, we make an additional (and optional) recommendation for using SED as complementary to the CEENE v2.0 (for the applicable LCI approaches).

For the midpoint assessment, ADP presented the higher final scores (Table 2) and thus is recommended as the midpoint LCIA method. Since it has different approaches (e.g., ultimate reserves), our recommendation is for the reserve base approach, as suggested in ILCD [3]. This is mainly due to the higher environmental relevance of that approach, *i.e.*, on one hand, the increase in resource scarcity leads to the exploitation of reserves that are less economic (e.g., marginal reserves), which are not accounted for in the economic reserve approach, and on the other hand, the ultimate reserve approach also includes reserves with a very low concentration, which may provide inconsistent results. This decision can be corroborated by the case study (Figure 4), in which the relevance of metals and minerals seemed to be negligible in the ultimate reserve approach.

For the endpoint assessment, the Recipe Endpoint and EPS2000 had the highest scores (Table 3), and are recommended as the endpoint LCIA method. Nevertheless, due to their low final score (3.5) in comparison to other LCIA methods from other approaches (CEENE v2.0 and ADP), this recommendation is made with limitations. Additionally, considering that the authors from the Recipe Endpoint, OGD and SuCo are from the same team (Radboud University and PRé Consultants), and also based on [49], we can assume that new versions of the Recipe Endpoint may include those new LCIA models. In that case, based on Tables 2 and 3 the score from the Recipe Endpoint in scientific robustness could increase. Thus, a study similar to this one should be performed again in the near future, in order to update the scores according to new versions that may come up.

2.4. Future Challenges and New Trends

The LCA scientific community has not yet reached consensus on how to evaluate resources. As previously mentioned, traditional approaches may be classified into three levels (Figure 1). However, there are new frameworks for the impact assessment of this category [6,7], showing that the LCIA methods available in literature are not yet consolidated, and new trends may appear in the future.

According to van Oers [36], resource depletion assessment at the midpoint and endpoint levels does not need to be regionalized. However, we cannot draw the same conclusion for RAM, where some type of resources may need to be assessed by spatial-differentiated CF, as done in CEENE v2.0 for land use (by LREx). Other RAM may follow the same trend, as ICEC/ECEC and SED for land use. Further, RAM may also need to regionalize other elementary flows, such as biotic resources and water. Regarding midpoint LCIA methods, we noticed an evolution, from the more traditional approach (EDIP), which assesses the availability of resources and, for the ADP, their extraction rate, for more recent methods that evaluate the decrease in the ore grade, as the Recipe Midpoint, ORI and OGD. Moreover, there are new trends that include the resources in the anthroposphere, for metals, as proposed in AADP. Currently, these new approaches are not totally suitable for LCA, due to the low number of CF, but this reality might change in the future (e.g., [42]). At the endpoint level, the assessment usually has higher uncertainties than at the midpoint level. However, newer LCIA models seem to be more scientifically robust, with fewer uncertainties (e.g., SuCo), and may be incorporated into traditional endpoint LCIA methods (e.g., Recipe Endpoint). Likewise, even though criticality does not yet have an operational LCIA method available, it is a well-established resource management methodology that may be incorporated into LCA in the future, especially when considering economic and social issues in LCSA [7,54].

Moreover, we could see that there are already a few operational LCIA methods with alternative approaches (outside of traditional LCA), for instance ICEC/ECEC, SED and Exergoecology, where concepts from Emergy and cradle-to-cradle (or grave-to-cradle) approaches, that have a high sustainability appeal, are brought into LCA.

3. Experimental Section

In order to search for different operational LCIA methods available in literature, we used different keyword combinations (e.g., resources and LCA) on web tools, such as Web of Science. Articles published from the last 20 years, until December of 2014, were considered. After that, the articles that referred to the LCIA methods *per se* were selected, *i.e.*, we excluded case studies that used those LCIA methods. Finally, we evaluated them through two criteria: (1) Scope: in which the amount of elementary flows that could be accounted for was evaluated, *i.e.*, the amount of CF available in the LCIA method. The availability of regionalized CF was also considered, but since spatial-differentiation in LCIA is not applicable to resource depletion assessment [36], this was considered solely for the RAM (not for midpoint and endpoint LCIA methods); (2) Scientific robustness: in which the model behind the LCIA method was evaluated, how it was scientifically proposed (theory used and/or cause-and-effect relation), how clear was the documentation, and if the method was fully operational. Other criteria could be used in our evaluation (e.g., acceptance by LCA community), but we preferred to focus on rather technical criteria. We gave scores between one (the lowest) and five (the highest) for each of these criteria and later we calculated an arithmetic average in order to provide a final score for each of the LCIA methods evaluated.

After the theoretical assessment of the LCIA methods, some of them were applied in a case study of ethylene production. For that, two scenarios were considered:

- A traditional FE, which was based on the data in the ecoinvent [55] dataset named "ethylene, average (RER) production, Alloc Def" (There is no dataset in ecoinvent for Brazilian ethylene). The inputs and outputs in this dataset are arranged as aggregated LCI; thus, it is not possible to clearly identify the life cycle stage of each elementary flow;
- A BE, from Brazil, where sugarcane is produced to generate ethanol, that is further dehydrated into ethylene. Therefore, Cavalett *et al.* [56] was used for sugarcane and ethanol data and the Swedish Life Cycle Center (CPM) database [57] for the ethanol-to-ethylene process unit. Sugarcane and ethanol production considered in reference [56] is from advanced technologic cultivation and production systems, from the state of São Paulo (Brazil). The ethanol-to-ethylene process unit is based on pilot scale data.

After modeling the life cycle of these two scenarios of ethylene, we performed an LCIA through different resource-based LCIA methods. Then, we evaluated which scenario had the best/worst results and searched for the main hotspots that each of these LCIA methods identified.

4. Conclusions

Through this study, we found 19 different LCIA methods for assessing the AoP Resources. We then made an assessment based on two criteria and, with support from a case study, were able to recommend the CEENE method v2.0 (with optional complementary assessment by SED) for process-based and hybrid LCI, and ICEC/ECEC for extended input-output LCI, at RAM level; ADP (reserve base) for midpoint level; and EPS2000 and the Recipe Endpoint for the endpoint level (with the possibility of only being the Recipe Endpoint in the future). In addition, it was possible to notice that the evaluation of the AoP Resources is not yet well established in the LCA community, not only due to the recent development of several new LCIA methods with different approaches, but also due to the new propositions on how this AoP should be assessed. In this sense, it is important to highlight the importance of performing a study similar to this one in the near future, as new LCIA methods and also new (and upgraded) versions of the current LCIA methods may appear.

Conflicts of Interest: The authors declare no conflict of interest.

References

1. MEA. *Ecosystem and Human Well-Being: The Millennium Ecosystem Assessment*; Island Press: Washington, DC, USA, 2015.
2. Swart, P.; Alvarenga, R.A.F.; Dewulf, J. Abiotic resource use. In *LCA Compendium—The Complete World of Life Cycle Assessment: Life Cycle Impact Assessment*, 1st ed.; Hauschild, M., Huijbregts, M.A.J., Eds.; Springer Press: Dordrecht, The Netherlands, 2015; Volume 4, pp. 247–269.
3. European Commission Joint Research Centre. *International Reference Life Cycle Data System (ILCD) Handbook—Recommendations for Life Cycle Assessment in the European Context*; Publications Office of the European Union: Luxembourg, Luxembourg, 2011.
4. Müller-Wenk, R. *Depletion of Abiotic Resources Weighted on the Base of 'Virtual' Impacts of Lower Grade Deposits in Future*; IWO Diskussionsbeitrag Nr. 57; Universität St. Gallen: St. Gallen, Switzerland, 1998.
5. Stewart, M.; Weidema, B.P. A consistent framework for assessing the impacts from resource use—A focus on resource functionality (8 pp). *Int. J. Life Cycle Assess.* **2005**, *10*, 240–247. [CrossRef]
6. Robech, J.T.; Vadenbo, C.; Hellweg, S.; Astrup, T.F. Impact assessment of abiotic resources in LCA: Quantitative comparison of selected characterization models. *Environ. Sci. Technol.* **2014**, *48*, 11072–11081. [CrossRef] [PubMed]
7. Dewulf, J.; Benini, L.; Mancini, L.; Sala, S.; Blengini, G.A.; Ardente, F.; Recchioni, M.; Maes, J.; Pant, R.; Pennington, D. Rethinking the area of protection "natural resources" in life cycle assessment. *Environ. Sci. Technol.* **2015**, *49*, 5310–5317. [CrossRef] [PubMed]
8. Liao, W.; Heijungs, R.; Huppes, G. Thermodynamic resource indicators in LCA: A case study on the titania produced in Panzhihua city, southwest China. *Int. J. Life Cycle Assess.* **2012**, *17*, 951–961. [CrossRef]
9. Dewulf, J.; Bosch, M.E.; Meester, B.D.; Vorst, G.V.D.; Langenhove, H.V.; Hellweg, S.; Huijbregts, M.A.J. Cumulative exergy extraction from the natural environment (CEENE): A comprehensive life cycle impact assessment method for resource accounting. *Environ. Sci. Technol.* **2007**, *41*, 8477–8483. [CrossRef] [PubMed]
10. Rugani, B.; Huijbregts, M.A.J.; Mutel, C.; Bastianoni, S.; Hellweg, S. Solar energy demand (SED) of commodity life cycles. *Environ. Sci. Technol.* **2011**, *45*, 5426–5433. [CrossRef] [PubMed]
11. Guinée, J. Development of a Methodology for the Environmental Life-Cycle Assessment of Products. PhD dissertation, Leiden University, Leiden, The Netherlands, 2 March 1995.
12. Life Cycle Impact Assessment Programme. Available online: http://www.lifecycleinitiative.org/activities/phase-i/life-cycle-impact-assessment-programme/ (accessed on 25 January 2016).
13. Hischier, R.; Weidema, B.; Althaus, H.-J.; Doka, G.; Dones, R.; Frischknecht, R.; Hellweg, S.; Humbert, S.; Jungbluth, N.; Loerincik, Y.; *et al.* Implementation of Life Cycle Impact Assessment Methods: Final Report Ecoinvent v2.1.; Swiss Centre for Life Cycle Inventories: St. Gallen, Switzerland, 2009; Volume 3.
14. Boustead, I.; Hancock, G.F. *Handbook of Industrial Energy Analysis*; Ellis Horwood Ltd.: New York, NY, USA, 1979.
15. Pimentel, D.; Hurd, L.E.; Bellotti, A.C.; Forster, M.J.; Oka, I.N.; Sholes, O.D.; Whitman, R.J. Food production and the energy crisis. *Science* **1973**, *182*, 443–449. [CrossRef] [PubMed]
16. VDI. *Cumulative Energy Demand—Terms, Definitions, Methods of Calculation*; VDI guideline 4600; Verein Deutscher Ingenieure: Dusseldorf, Germany, 1997.
17. Frischknecht, R.; Wyss, F.; Knöpfel, S.B.; Lützkendorf, T.; Balouktsi, M. Cumulative energy demand in LCA: The energy harvested approach. *Int. J. Life Cycle Assess.* **2015**, *20*, 957–969. [CrossRef]

Resources **2016**, 5, 13

18. Huijbregts, M.A.J.; Rombouts, L.J.A.; Hellweg, S.; Frischknecht, R.; Hendriks, A.J.; van de Meent, D.; Ragas, A.M.J.; Reijnders, L.; Struijs, J. Is cumulative fossil energy demand a useful indicator for the environmental performance of products? *Environ. Sci. Technol.* **2006**, *40*, 641–648. [CrossRef] [PubMed]
19. Huijbregts, M.A.J.; Hellweg, S.; Frischknecht, R.; Hendriks, H.W.M.; Hungerbuhler, K.; Hendriks, A.J. Cumulative energy demand as predictor for the environmental burden of commodity production. *Environ. Sci. Technol.* **2010**, *44*, 2189–2196. [CrossRef] [PubMed]
20. Alvarenga, R.A.F. Environmental Sustainability of Biobased Products: New Assessment Methods and Case Studies. PhD Dissertation, Ghent University, Ghent, Belgium, 13 June 2013.
21. Alvarenga, R.A.F.; Dewulf, J.; Langenhove, H.; Huijbregts, M.A.J. Exergy-based accounting for land as a natural resource in life cycle assessment. *Int. J. Life Cycle Assess.* **2013**, *18*, 939–947. [CrossRef]
22. Bösch, M.; Hellweg, S.; Huijbregts, M.; Frischknecht, R. Applying cumulative exergy demand (CExD) indicators to the ecoinvent database. *Int. J. Life Cycle Assess.* **2007**, *12*, 181–190. [CrossRef]
23. Dewulf, J.; Van Langenhove, H.; Muys, B.; Bruers, S.; Bakshi, B.R.; Grubb, G.F.; Paulus, D.M.; Sciubba, E. Exergy: Its potential and limitations in environmental science and technology. *Environ. Sci. Technol.* **2008**, *42*, 2221–2232. [CrossRef] [PubMed]
24. Odum, H.T. *Environmental Accounting: Emergy and Environmental Decision Making*, 1st ed.; John Wiley & Sons: New York, NY, USA, 1996.
25. Rugani, B.; Benetto, E. Improvements to emergy evaluations by using life cycle assessment. *Environ. Sci. Technol.* **2012**, *46*, 4701–4712. [CrossRef] [PubMed]
26. Ingwersen, W. Emergy as a life cycle impact assessment indicator. *J. Ind. Ecol.* **2011**, *15*, 550–567. [CrossRef]
27. Schmidt-Bleek, F. *The Fossil Makers*; Birkhäuser: Basel, Boston, Berlin, 1993.
28. Ritthoff, M.; Rohn, H.; Liedtke, C. *MIPS BErechnen: Ressourcen Produktivität von Produkten und Dienstleistungen*; Wuppertal Spezial, Wuppertal Institutfür Klima, Umwelt und Energie 27; Visualisation Lab Wuppertal Institut: Wuppertal, Germany, 2002.
29. Saurat, M.; Ritthoff, M. Calculating MIPS 2.0. *Resources* **2013**, *2*, 581–607. [CrossRef]
30. Wiesen, K.; Saurat, M.; Lettenmeier, M. Calculating the material input per service unit using the ecoinvent database. *Int. J. Perform. Eng.* **2014**, *10*, 357–366.
31. Hau, J.L.; Bakshi, B.R. Expanding exergy analysis to account for ecosystem products and services. *Environ. Sci. Technol.* **2004**, *38*, 3768–3777. [CrossRef] [PubMed]
32. Zhang, Y.; Baral, A.; Bakshi, B.R. Accounting for ecosystem services in life cycle assessment, Part II: Toward an ecologically based LCA. *Environ. Sci. Technol.* **2010**, *44*, 2624–2631. [CrossRef] [PubMed]
33. Szargut, J.; Morris, D.R.; Steward, F.R. *Exergy Analysis of Thermal, Chemical, and Metallurgical Processes*; Springer: Berlin, Germany, 1998.
34. Wackernagel, M.; Rees, W. *Our Ecological Footprint: Reducing Human Impact on the Earth*; NSP: Gabriola Island, BC, Canada, 1998.
35. Huijbregts, M.A.J.; Hellweg, S.; Frischknecht, R.; Hungerbuhler, K.; Hendriks, A.J. Ecological footprint accounting in the life cycle assessment of products. *Ecol. Econ.* **2008**, *64*, 798–807. [CrossRef]
36. van Oers, L.; de Koning, A.; Guinee, J.; Huppes, G. *Abiotic Resource Depletion in LCA—Improving Characterization Factors for Abiotic Resource Depletion as Recommended in the New Dutch LCA Handbook*; Road and Hydraulic Engineering Institute: Leiden, The Netherlands, 2002.
37. Hauschild, M.; Wenzel, H. *Environmental Assessment of Products: Scientific background*; Chapman & Hall: London, UK, 1998; Volume 2.
38. Klinglmair, M.; Sala, S.; Brandão, M. Assessing resource depletion in LCA: A review of methods and methodological issues. *Int. J. Life Cycle Assess.* **2014**, *18*, 580–592. [CrossRef]
39. Goedkoop, M.; Heijungs, R.; Huijbregts, M.; de Schryver, A.; Struijs, J.; van Zelm, R. *ReCiPe 2008—A life Cycle Impact Assessment Method which Comprises Harmonized Category Indicators at the Midpoint and the Endpoint Level*, 1st ed.; Report I: Charaterisation; Ministry of Housing, Spatial Planning and the Environment (VROM): The Hague, The Netherlands, 2009.
40. Swart, P.; Dewulf, J. Quantifying the impacts of primary metal resource use in life cycle assessment based on recent mining data. *Resour. Conserv. Recycl.* **2013**, *73*, 180–187. [CrossRef]
41. Schneider, L.; Berger, M.; Finkbeiner, M. The anthropogenic stock extended abiotic depletion potential (AADP) as a new parameterisation to model the depletion of abiotic resources. *Int. J. Life Cycle Assess.* **2011**, *16*, 929–936. [CrossRef]

42. Schneider, L.; Berger, M.; Finkbeiner, M. Abiotic resource depletion in LCA—Background and update of the antropogenic stock extended abiotic depletion potential (AADP) model. *Int. J. Life Cycle Assess.* **2015**, *20*, 709–721. [CrossRef]

43. Vieira, M.D.M.; Goedkoop, M.J.; Storm, P.; Huijbregts, M.A.J. Ore grade decrease as life cycle impact indicator for metal scarcity: The case of copper. *Environ. Sci. Technol.* **2012**, *46*, 12772–12778.

44. Goedkoop, M.; Spriensma, R. *The Eco-Indicator 99—A Damage Oriented Method for Life Cycle Impact Assessment: Methodology Report*; PRé Consultants: Amersfoort, The Netherlands, 2000.

45. Chapman, P.F.; Roberts, F. *Metal Resources and Energy*; Butterworths Monographs in Materials: London, UK, 1983.

46. Ponsioen, T.C.; Vieira, M.D.M.; Goedkoop, M.J. Surplus cost as a life cycle impact indicator for fossil resource scarcity. *Int. J. Life Cycle Assess.* **2014**, *19*, 872–881. [CrossRef]

47. Steen, B. *A Systematic Approach to Environmental Priority Strategies in Product Development (EPD). Version 2000—General System Characteristics*; CPM report. Nr. 4; Centre for Environmental Assessment of Products and Material Systems, Chalmers University of Technology, Technical Environmental Planning. Chalmers University of Technology: Gothenburg, Sweden, 1999.

48. Steen, B. *A Systematic Approach to Environmental Priority Strategies in Product Development (EPD). Version 2000—Models and Data of the Default Method*; CPM report. Nr. 5; Centre for Environmental Assessment of Products and Material Systems, Chalmers University of Technology, Technical Environmental Planning. Chalmers University of Technology: Gothenburg, Sweden, 1999.

49. Ponsioen, T.; PRé Consultants, Amersfoort, The Netherlands. Personal communication. 7 August 2015.

50. Valero, A.; Valero, A. Exergoecology: A thermodynamic approach for accounting the earth's mineral capital. *The case of bauxite-aluminium and limestone-lime chains. Energy* **2010**, *35*, 229–238.

51. Valero, A.; Valero, A. From grave to cradle. *J. Ind. Ecol.* **2012**, *17*, 43–52. [CrossRef]

52. Frischknecht, R.; Steiner, R.; Jungbluth, N. *The Ecological Scarcity Method—Eco-Factors 2006. A Method for Impact Assessment in LCA*; Bundesamtfür Umwelt (BAFU): Bern, Switzerland, 2009.

53. Berger, M.; Finkbeiner, M. Correlation analysis of life cycle impact assessment indicators measuring resource use. *Int. J. Life Cycle Assess.* **2011**, *16*, 74–81. [CrossRef]

54. Vandenbo, C.; Rorbech, J.; Haupt, M.; Frischknecht, R. Abiotic resources: New impact assessment approaches in view of resource efficiency and resource criticality. *Int. J. Life Cycle Assess.* **2014**, *19*, 1686–1692.

55. Ecoinvent. *Ecoinvent Data v3.0*; Swiss Centre for Life Cycle Inventories: Dübendorf, Switzerland, 2015.

56. Cavalett, O.; Chagas, M.; Seabra, J.; Bonomi, A. Comparative LCA of ethanol versus gasoline in Brazil using different LCIA methods. *Int. J. Life Cycle Assess.* **2012**, *18*, 647–658. [CrossRef]

57. CPM. *Center for Environmental Assessment of Product and Material Systems (CPM) LCA Database*; CPM Consortium: Gothenburg, Sweden, 2008.

resources

MDPI

Article

Exergy as a Measure of Resource Use in Life Cyclet Assessment and Other Sustainability Assessment Tools

Goran Finnveden [1,*], Yevgeniya Arushanyan [1] and Miguel Brandão [1,2]

[1] Department of Sustainable Development, Environmental Science and Engineering (SEED), KTH Royal Institute of Technology, Stockholm SE 100-44, Sweden; yaru@kth.se (Y.A.); miguel.brandao@abe.kth.se (M.B.)

[2] Department of Bioeconomy and Systems Analysis, Institute of Soil Science and Plant Cultivation, Czartoryskich 8 Str., 24-100 Pulawy, Poland

* Correspondance: goran.finnveden@abe.kth.se; Tel.: +46-8-790-73-18

Academic Editor: Mario Schmidt
Received: 14 December 2015; Accepted: 12 June 2016; Published: 29 June 2016

Abstract: A thermodynamic approach based on exergy use has been suggested as a measure for the use of resources in Life Cycle Assessment and other sustainability assessment methods. It is a relevant approach since it can capture energy resources, as well as metal ores and other materials that have a chemical exergy expressed in the same units. The aim of this paper is to illustrate the use of the thermodynamic approach in case studies and to compare the results with other approaches, and thus contribute to the discussion of how to measure resource use. The two case studies are the recycling of ferrous waste and the production and use of a laptop. The results show that the different methods produce strikingly different results when applied to case studies, which indicates the need to further discuss methods for assessing resource use. The study also demonstrates the feasibility of the thermodynamic approach. It identifies the importance of both energy resources, as well as metals. We argue that the thermodynamic approach is developed from a solid scientific basis and produces results that are relevant for decision-making. The exergy approach captures most resources that are considered important by other methods. Furthermore, the composition of the ores is shown to have an influence on the results. The thermodynamic approach could also be further developed for assessing a broader range of biotic and abiotic resources, including land and water.

Keywords: Life Cycle Assessment; resource use; exergy; waste; recycling; metals

---

1. Introduction

A more efficient utilization of resources, from, e.g., improved waste treatment, is an important part of a more circular economy, which is arguably a prerequisite for a sustainable society. Life Cycle Thinking (LCT) and Life Cycle Assessment (LCA) have received prominent positions in European waste policy, for example in the Waste Framework Directive, as an approach and tool, respectively, for assessing the environmental impacts and resource use associated with alternative waste management strategies [1].

LCA is a tool to assess the potential environmental impacts and resources used throughout a product's life cycle, i.e., from raw material extraction, via production and use phases, to waste management [2–4]. The term "product" includes both goods and services, for example waste management. The unique feature of LCA is the focus on a life-cycle perspective. This implies that system boundaries should be so wide that inputs to the life-cycle system should be resources as we find them in nature, and outputs should be emissions to nature.

There are four phases in an LCA study: Goal and Scope Definition, Life Cycle Inventory Analysis (LCI), Life Cycle Impact Assessment (LCIA), and Interpretation. The result from the LCI is a compilation of the inputs (resources) and the outputs (emissions) from the product over its life cycle. The ISO standards [4] further define Life Cycle Impact Assessment (LCIA) as the phase of an LCA study that involves classification and characterization of substances, and optionally normalization and weighting. In the classification the LCI, results on resources and emissions are classified into impact categories of environmental relevance (e.g., kg CO_2 emissions into Global Warming). In the characterization part of the LCIA, the contributions of different inputs and outputs to impact categories are modeled quantitatively and expressed as an impact score in a unit common to all contributions within the impact category [2]. The relative contribution of a substance to an impact category is estimated with characterization models (e.g., 1 kg of CH_4 contributes around 30 times as much to cumulative radiating forcing as 1 kg of CO_2). Normalization and weighting are optionally used in order to enable comparison of results across impacts (e.g., global warming against resource depletion).

The ISO standard for LCA [4] states that "resource use" is one of the categories of environmental impacts needing consideration. In the LCA literature, "resource depletion" (including minerals, fossil and renewable energy resources and water) is often described as one impact category that should be included in an LCA study. In principle, the terms "resource use", "resource consumption" and "resource depletion" should mean different things (resource use can lead to consumption (implying destruction of the resource) which could lead to depletion of the stock or fund resource). In practice, however, many scholars and practitioners are using the terms interchangeably, leading to a mixed use of the terms.

The characterization of the use of abiotic resources—such as minerals and non-renewable fuels—is one of the most frequently discussed impact categories and consequently there is a wide variety of methods available for characterizing contributions to this category (e.g., [2,5–8]). Essentially, four different groups of approaches can be distinguished [9,10]:

1. The impact of the present use of resources is modeled as that of the future use of resources. The rationale for this approach is that the current use of non-renewable resources implies that more effort (e.g., energy) to extract the same amount will be required by future generations, assuming that ore grades decrease with greater extraction, and that technology remains the same. Alternatively, we can adopt other resources that substitute for the ones being assessed as proxies, so that current impacts can be measured in terms of the future impacts. This has been the basis for several LCIA methods for resources, for example Eco-indicator 99 [11] and ReCiPe [12]. However, it has been argued that if current resource use leads to changes in the environmental interventions of future extractions, this should be modeled in the Inventory Analysis, and not in the LCIA [2,13].

2. There are methods related to some measure of available resources or reserves and extraction rates. Different approaches exist based on different measures of the reserves, e.g., technically and economically available reserves [14] or ultimately available reserves, as in the CML approach [15,16], and extraction rates. For example, the CML approach uses antimonium (Sb) as the reference of resource extraction rate–to-reserve ratio, such that:

$$ADP_i = \frac{DR_i/(R_i)^2}{DR_{sb}/(R_{sb})^2} \tag{1}$$

where [15]:

ADP_i Abiotic Depletion Potential of resource i (dimensionless);
R_i ultimate reserve of resource i (kg);
DR_i extraction rate of resource i $(kg \cdot y^{-1})$
R_{sb} ultimate reserve of antimonium (i.e., the reference resource (kg));
D_{sb} extraction rate of antimonium R_{sb} (i.e., the reference resource $(kg \cdot y^{-1})$).

3. There are thermodynamic methods based on exergy decrease or entropy increase. Exergy is a measure of available energy. Entropy can be interpreted in many ways, e.g., as a measure of disorder but also as a measure for the dispersal of energy. In contrast to energy, exergy is destroyed in all real world processes as entropy is produced [17]. The exergy decrease and the entropy increase mirror each other. Methods and data based on this approach have been developed for LCA [18–23]. The exergy content of the resources can be assumed to be used through incineration, other chemical reactions and dissipation [18]. Although it may be claimed that all the approaches rely on thermodynamics, this group will be called the thermodynamic approach in this paper.

4. The last approach is the aggregation of the total use of energy as in the Cumulative Energy Demand (CED) [24].

All LCIA methods across all approaches above have in common that they result in characterization factors CF_i that can be used for calculating the indicator for the use of abiotic resources according to Equation (2) (e.g., [3,9]):

$$I = \Sigma CF_i \, m_i \tag{2}$$

where I is the indicator result and m_i is the use of the abiotic resource i in the system under study.

The aim of this paper is to contribute to the discussion on different approaches to measure the use of abiotic natural resources (such as non-renewable fuels and mineral resources) in LCA and other methods. This is done by illustrating the application of the thermodynamic approach to case studies, and by comparing the results with other approaches. The focus is on the thermodynamic approach which is discussed in more detail.

2. Materials and Methods

Two case studies were chosen to illustrate the use of different methods: one on the recycling of ferrous waste and one on the production and use of a laptop computer. Since one purpose of recycling is to save resources, a simple recycling case was chosen to see if the methods would capture this concern. A laptop was chosen to illustrate a more complex product, which includes many different types of metals and plastics made from fossil fuels, and which uses significant amounts of electricity during the use phase. For both cases, data were taken from the Ecoinvent database [25] as implemented in the SimaPro software (Pré, Amersfoort, the Netherlands, www.pre.nl).

2.1. Ferrous Waste Recycling

This case is based on the Ecoinvent 2 dataset "Steel, electric, un- and low-alloyed, at plant/RER" [26] with the addition of avoided virgin production (1 kg of steel, low-alloyed per every 1.1 kg of material recycled), with removal of steel scrap input. Direct use of electricity has been replaced by the Swedish electricity mix [27]. The system studied represents an average of different processes of secondary steel production with EU technology and includes transportation of scrap metal and other input materials to the electric arc furnace, steel-making process and casting.

2.2. Production and Use of a Laptop

The case is based on the Ecoinvent 2 dataset "Use, computer, laptop, office use/RER" [28]. The system studied includes laptop production and disposal, transportation to the user and energy consumption during use, the latter considered over a lifespan of four years and taking into account different use modes for office use. The functional unit of the assessment was 1 laptop.

The process for laptop production includes raw material extraction and processing, manufacturing of all components, energy and water use required for various processes, transportation for input materials (ship, rail and road), packaging, final disposal of the laptop, and infrastructure required (factory). The dataset is global and describes the manufacturing of a typical laptop computer during the years 2002–2005. The final disposal is represented by a mixture of manual and mechanical WEEE treatment in Switzerland. The benefit of material recovery is not accounted for. Rail and road transportation is used for the delivery to a user. For the use phase Swedish electricity is used [27]. The office use is calculated according to the share between active (5.5 h/day), standby (2 h/day) and off (16.5 h/day) modes.

2.3. Characterization Methods

Two versions of the thermodynamic approach were applied here. The first is called Exergy F & Ö, where the exergy data were taken from [18]. These data were calculated from information on the chemical composition of the material, the basic thermodynamic data and the reference state developed by Szargut et al. [29] which is made to be similar to the natural environment, including common components of the atmosphere, sea and crust of the earth. For fossil fuels, the exergy is close to the lower heating value. For metals, the whole metal ore was considered as an input to the technical system in line with the life-cycle approach [18]. The second version of the thermodynamic approach was the Cumulative Exergy Demand (CExD) method [19], which uses other data for the ore compositions (see below) and an updated version of the reference state including, for example, more recent data on concentrations of trace elements in sea water (the reader is referred to the references for more details). De Meester et al. [30] have shown that an updated reference state in most cases has only a minor influence on the results. For some specific minerals, the differences can, however, be larger.

In addition, several other often-used LCIA methods were included, e.g., Cumulative Energy Demand (CEnD), CML, Eco-indicator 99 and ReCiPe for minerals and fossil fuels, all as implemented in Simapro. All of these are intended to be used as methods for characterization of the use of abiotic resources in LCA and are in that sense comparable. Although different terms may be used in the literature when describing the methods (e.g., resource use, resource consumption or resource depletion), they are intended and used for this same purpose. All use information about the used amounts of the abiotic resources from the LCIA. The system boundaries for the Ecoindicator 99 and ReCiPe methods are expanded compared to the other methods since they include also future resource extraction. This is because information on impacts from future resource extraction is used for the indicator on the impacts of the current resource use. The methods also differ in the age of underlying data (the readers are referred to the specific methods for details). It is, however, likely that the difference in age has a minor influence in the comparisons since the differences in approaches are large and large parts of the data (e.g., on thermodynamic properties) are not expected to change significantly with time. Many of them are also used on a routine basis by LCA practitioners.

3. Results and Discussion

The results for ferrous recycling are presented in Table 1 and Figure 1. The results are in most cases negative, indicating that recycling saves resources. The exception is uranium, which is used for the production of nuclear power. The results thus show that although the total use of abiotic resources decreases, the electricity use is increased.

The results for the production and use of a laptop are presented in Table 2 and Figure 2.

Table 1. Characterization of most important non-renewable resources for ferrous recycling in percentage of the total result (only resources that contribute to more than 1% of total result according to at least one method are included; "0" means less than 0.1% and n/a means that no characterization factor for this input was available).

Resources	Exergy F & Ö	CExD	CEnD	CML	Eco-Indicator 99	ReCiPe Mineral	ReCiPe Fossil Fuels
Coal	−46	−64	−88	−88	−6	n/a	−84
Gas	−3	−5	−8	−5	−9	n/a	−8
Oil	−4	−7	−9	−6	−10	n/a	−9
U	2	3	4	0	n/a	0	n/a
Ni	−48	−15	n/a	−0.1	−70	−20	n/a
Mo	n/a	−2	n/a	−0.1	−1	−3	n/a
Fe	−2	−9	n/a	0	−2	−25	n/a
Mn	n/a	−0.3	n/a	0	−0.2	−35	n/a
Cr	−0.1	−0.5	n/a	0	−1	−16	n/a

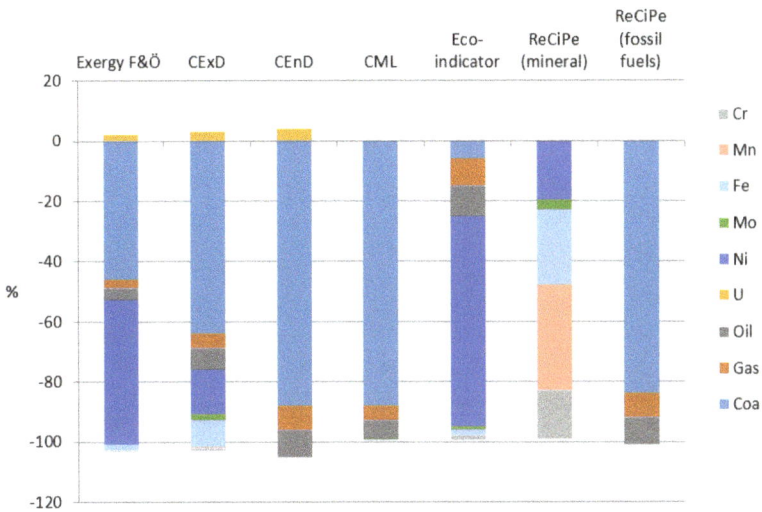

Figure 1. Characterization of most important non-renewable resources for ferrous recycling in percentage of the total result.

Table 2. Characterization of most important non-renewable resources for production and use of a laptop in percentage of the total result (only resources that contribute to more than 1% of total result according to at least one method are included; "0" means less than 0.1% and n/a means that no characterization factor for this input was available).

Resources	Exergy F & Ö	CExD	CEnD	CML	Eco-Indicator	ReCiPe Minerals	ReCiPe Fossil Fuels
Coal	12	17	18	45	2	n/a	39
Gas	9	13	15	27	41	n/a	33
Oil	6	12	13	20	34	n/a	27
U	24	47	53	0	n/a	0.3	n/a
Cu	6	1	n/a	0.1	9	16	n/a
Au	41	8	n/a	7	n/a	49	n/a
Ni	3	0.8	n/a	0	n/a	0.1	n/a
Mo	n/a	0.1	n/a	0	0.2	1.3	n/a
Fe	0	0.1	n/a	0	0.1	1.5	n/a
Mn	n/a	0	n/a	0	0	3	n/a
Cr	0	0	n/a	0	0.1	5	n/a
Sn	n/a	0.3	n/a	0.1	6	18	n/a

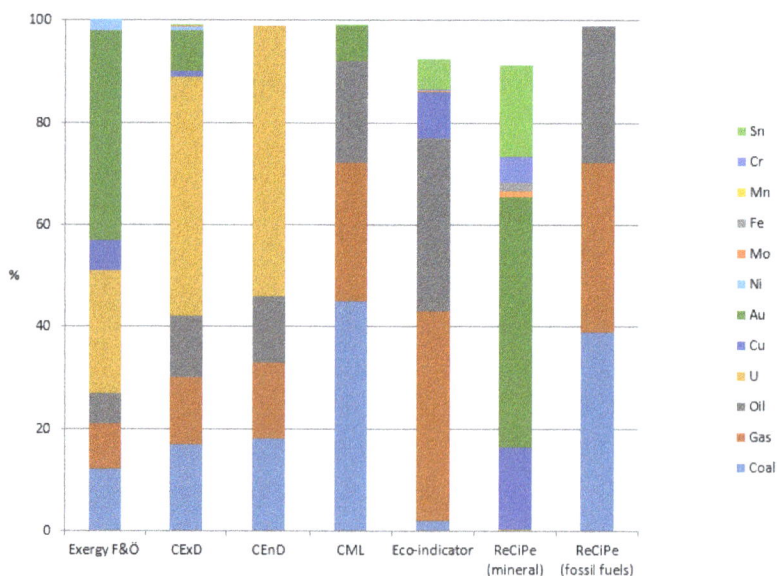

Figure 2. Characterization of most important non-renewable resources for the production and use of a laptop in percentage of the total result.

The Exergy F & Ö method and CExD give similar results for traditional energy resources, but partly different results for metals. The CEnD naturally only gives results for energy resources, and the CML method gives the result that energy resources are, in practice, the most important resources, especially in the case of ferrous recycling. Conversely, Eco-indicator 99 suggests that metals are more important than energy resources in this case. The Exergy F & Ö and CExD methods give weight to both energy and metal resources.

For metals, the Exergy F & Ö and CExD methods give different results. The Exergy F & Ö method gives higher results than CExD for some metals (e.g., Ni in the case of ferrous recycling and Au in the laptop case). This is because the two methods have modeled the metal ores in different ways. The Exergy F & Ö method has data for sulphidic ores for some metals (copper, gold and nickel) with a high chemical exergy whereas the CExD method uses a generic average ore matrix with a lower chemical exergy. The exergy of the generic ore used in the CExD method [19] was the average of the ores in [18]. It can therefore be argued that the ore composition adopted in the Exergy F & Ö method is more specific and relevant for the metal in question. On the other hand, the ore-generic data may be considered more generally applicable. The ore-specific data used in the Exergy F & Ö method can possibly be compared with site-dependent data, which is sometimes used in LCIA (cf. [2]). However, both methods are consistent in identifying the same metals as being the most important ones. This is partly in contrast to the Ecoindicator 99 and ReCiPe methods, which highlight other metals, partly because of data gaps. It also interesting to note that U is highlighted in the laptop case by the CEnD, CExD and the Exergy F & Ö methods, but hardly at all by the other methods.

Since different methods produce different answers, it is important to discuss and evaluate different approaches. Important questions then are: How can a choice be made between different methods and when is a specific method more adequate? Unfortunately, there is no simple way of determining this. There is no method by which it can be shown that one characterization method is the "correct" one (cf. [31]). Instead, theoretical reasoning must be applied and underlying assumptions should be discussed. Examples of questions that can be raised include: Is the problem defined in a relevant way? Does the quantification method adequately and reliably quantify the contribution to the problem?

Are there any logical contradictions? Does the method produce reasonable results and can we in any sense judge which results are reasonable?

In the thermodynamic approach, resource use is measured in terms of exergy use (indicated by the exergy content of the resources) or entropy production. Several ways of thinking can lead to the conclusion that exergy is a relevant measure:

- The energetic argument claims that useful energy (i.e., exergy) is the ultimate limiting and scarce resource because every material resource has an energy cost associated with extracting it, which limits its scale [32]. Given sufficient amounts, a society can divert exergy, within the current technical possibilities, to the acquisition of whatever material that is in short supply. Exergy, rather than energy, is used in this context since consideration is given to the quality of the energy (i.e., the ability to do work) as well as the chemical exergy of traditionally non-energetic raw materials [18]. As ore grades become lower, more exergy will be needed to extract the resource. This is for two reasons. One is because more energy is needed to extract larger amounts of ore (which will have lower grade/concentrations of the resource). The other one is because the amount of chemical exergy that comes with the ore will increase (assuming a constant amount of exergy per kg of ore) since more ore will be needed to produce a certain product.

- The usability argument starts with the question: When we discuss resource depletion or consumption, what is actually depleted or consumed? [18]. It is neither matter, since matter cannot be destroyed or consumed (except for nuclear reactions), nor energy. A reasonable answer may be that it is the usable energy and matter that is consumed and depleted and transformed to less usable energy and matter. This happens in all real world processes. It could therefore be relevant to have the use of usable energy and matter as an indicator for resource consumption. A measure of useful energy is exergy. For a material to be useful it must normally be ordered, i.e., structured and concentrated. A well-known scientific measure that is often interpreted as a measure of the disorder of a system is entropy. If a material is to be useful, it must normally have lower entropy than the surroundings, here defined as the reference state. This implies that the material is structured and concentrated, i.e., has a higher order. An example is an ore which has a higher concentration and typically a lower entropy than the average crust defined in the reference state used in the methods applied here. Another example is fresh water which has lower entropy than seawater (which is part of the reference state). Societies and technical systems can be described as systems feeding on low-entropy matter and energy and converting them into high-entropy matter and energy. The entropy production may therefore be considered a relevant measure of resource consumption (cf. [33]). Since entropy is increased as exergy is decreased according to Ek 3 (where δE is the decrease of exergy, T_0 is the temperature of the surroundings and $\Sigma \Delta S$ is the sum of the entropy increased), exergy can be used instead of entropy [29].

$$\delta E = T_0 \sum \Delta S \qquad (3)$$

- The reversibility argument starts with the assumption that a relevant measure of resource consumption may be the costs required to restore the resource. The exergy of a resource is the theoretical minimum energy required to produce this resource from the defined reference state. It may thus be seen as measure of the minimum costs associated with the resource.

Are the results provided by the thermodynamic method reasonable? When applied in practice, it can be noted that the thermodynamic approach as used here can highlight both energetic resources (both fossil fuels and nuclear fuels) as well as metal resources. This is in contrast to some of the other methods which either only include energetic or metal resources (such as ReCiPe) or in practice highlight only one of them (such as the CML method) or have significant data gaps (such as Ecoindicator). We believe that that is an indication that the results are relevant to decision-makers and useful. This is because both energy and metal resources are on the policy agenda. The method can thus provide support for decision-making when, for example, different products using different types of resources are compared.

However, despite exergy being a feature that is common to both energy and non-energy resources, it is limited in capturing the scarcity of a material, as no indication of abundance and extraction rates of the material is included therein (e.g., [5,10]). Instead the thermodynamic approach is based on the assumption that different resources can substitute for each other and that the limited and scarce resource is exergy. So instead of capturing the scarcity of materials, it is focused on the scarcity of exergy. This is relevant since physical scarcity of materials can be overcome with exergy. If there is enough exergy, materials can be extracted from low-grade ores and seawater.

The thermodynamic approach (or the other methods discussed here) does, however, not capture all resource aspects that are on the political agenda. For example, geopolitical aspects, which are a part of resource criticality assessments (e.g., [34]), are not included. However, there is currently no common and clear picture of how to describe the problem of resource use and which aspects should be included and valued (cf. [35]). Since different methods for the characterization of abiotic resources capture different aspects, they can also be seen as complementary.

The thermodynamic approach can be further developed. Different types of metal ores have different exergies [18] and the databases should therefore be improved to reflect this variability. This is especially relevant if changing ore grades also leads to changes in the composition of the ores. This is the case for many metals typically found in sulphide ores at higher ore grades (e.g., copper), but in oxide ores at lower ore grades with corresponding changes in the exergy of the ore [18]. The possibilities of using exergy consumption not only as measure of abiotic resources but also for other types of resources such as biotic resources, water and land [20] is also an interesting development. This could involve further complexities related, for example, to time for replenishing fund resources, local scarcity of water, etc. The advantage is, however, the possibility to integrate different types of resources in a common framework.

Although characterization methods for the use of abiotic resources have been discussed extensively, there are only a limited number of published studies where comparisons including the thermodynamic approach were made (notable exceptions include [6–8] but these were limited in different ways). The comparisons made have also focused on the Cad data [19] and not the Exergy F & Ö data [18], although we show here that they can give different results. Since different case studies illustrate different aspects, new learnings can be made and further studies would be useful. In addition to showing the differences between alternative methods, the case studies here illustrate that the thermodynamic approach can show the importance of both energy and material resources, and that the two datasets used here give similar but not identical results illustrating the importance of the ore composition.

LCA and exergy analysis can be linked in many ways, (cf. [21]). Here, the focus has been on exergy as measure of resource use in the characterization part of the LCIA. In a number of case studies (e.g., [36–38]), exergy analysis is used as a characterization method or as a method for broadening LCA [39]. It can be used in these ways for both attributional (or accounting) LCA, as well as consequential LCA (cf. [2]). Others have suggested that LCA and exergy analysis complement each other (e.g., [40–43]. Exergy analysis can also be used as a stand-alone assessment tool, e.g., [29,44–47], or linked to other environmental systems analysis tools, such as Strategic Environmental Assessment [48], besides its use as an engineering tool for analyzing and optimizing different types of processes [29,49]. This non-exhaustive list indicates that the applications of exergy analysis are numerous and that it is an established method in many areas. It could therefore also be used with confidence as a characterization method in LCA.

4. Conclusions

There are a number of different methods available for characterizing non-renewable resources and they give strikingly different results. The thermodynamic approach is developed from a solid scientific basis and it is currently operational with, for example, the CExD method with an extensive database. The approach is used by different groups and it produces relevant results. It can be noted that it captures most abiotic resources that are considered important by other methods. For the case studies on the recycling of ferrous waste and the production and use of a laptop, both energy resources and metals are of importance. Results for the thermodynamic approach depend on the composition of the ore; more ore-specific data should be developed and their importance evaluated (in parallel to site-specific data for emissions and impacts). Possibilities for using thermodynamic data for other resources (renewable materials and energy, water and land) should also be explored further.

Revisiting the questions raised above, we draw the conclusions that the thermodynamic approach used here is developed from an established scientific basis, based on a discussion on why exergy is a relevant measure. It is relevant as an indicator both for resource use and resource depletion since it is a measure of something that is actually consumed. We also argue that the results are relevant for decision-making. The calculations can be made in a reliable way and the data necessary is robust and time-independent, except for the composition of the ores which may change with time as the ore grades and types of ores used change.

Acknowledgments: Funding from Vinnova and partners of CESC (Center for Sustainable Communication) at KTH is appreciated. Comments from anonymous reviewers are also gratefully acknowledged.

Author Contributions: G.F. designed the research, Y.A. made the calculations, G.F. drafted the paper with contributions from Y.A. and M.B.

Conflicts of Interest: The authors declare no conflict of interest.

References

1. Lazarevic, D.; Buclet, N.; Brandt, N. The application of life cycle thinking in the context of European waste policy. *J. Clean. Product.* **2012**, *29–30*, 199–207. [CrossRef]
2. Finnveden, G.; Hauschild, M.; Ekvall, T.; Guinée, J.; Heijungs, R.; Hellweg, S.; Koehler, A.; Pennington, D.; Suh, S. Recent developments in Life Cycle Assessment. *J. Environ. Manag.* **2009**, *91*, 1–21. [CrossRef] [PubMed]
3. Curran, M.A., Ed.; *Life Cycle Assessment Student Handbook*; Wiley-Scivener: Hoboken, NJ, USA; Salen, MA, USA, 2015.
4. *Environmental Management—Life Cycle Assessment—Principles and Framework*; ISO 14040:2006; ISO Standard: Geneva, Switzerland, 1997.
5. Klinglmair, M.; Sala, S.; Brandão, M. Assessing resource depletion in LCA: A review of methods and methodological issues. *Int. J. Life Cycle Assess.* **2014**, *19*, 580–592. [CrossRef]
6. Rørbech, J.T.; Vadenbo, C.; Hellweg, S.; Aastrup, T.F. Impact assessment of abiotic resources in LCA: Quantitative comparison of selected characterization models. *Environ. Sci. Technol.* **2014**, *48*, 11072–11081. [CrossRef] [PubMed]
7. Van Caneghem, J.; Vermeulen, I.; Block, C.; Cramm, P.; Mortier, R.; Vandercasteele, C. Abiotic depletion due to resource consumption in a steelwork assessed by five different methods. *Resour. Conserv. Recycl.* **2010**, *54*, 1067–1073. [CrossRef]
8. Alvarenga, R.A.F.; de Oliveira Lins, I.; de Almeida Net, J.A. Evaluation of abiotic resource LCIA methods. *Resources* **2016**, *5*. [CrossRef]
9. Finnveden, G. "Resources" and related impact categories. In *Towards a Methodology for Life Cycle Impact Assessment*; Udo de Haes, H.A., Ed.; SETAC-Europe: Brussels, Belgium, 1996; pp. 39–48.
10. European Commission-Joint Research Centre—Institute for Environment and Sustainability. *International Reference Life Cycle Data System (ILCD) Handbook-Recommendations for Life Cycle Impact Assessment in the European Context*, 1st ed.; EUR 24571 EN; Publications Office of the European Union: Luxemburg, Luxemburg, 2011.
11. Goedkoop, M.; Spriensma, R. *The Eco-indicator 99—A Damage-oriented Method for Life Cycle Impact Assessment. Methodology Report*, 2nd ed.; Pré Consultants, B.V.: Amersfoort, The Netherlands, 2000.

12. Goedkoop, M.J.; Heijungs, R.; Huijbregts, M.; De Schryver, A.; Struijs, J.; Van Zelm, R. ReCiPe 2008, A life Cycle Impact Assessment Method Which Comprises Harmonised Category Indicators at the Midpoint and the Endpoint Level, First edition Report I: Characterisation. Available online: http://www.lcia-recipe.net (accessed on 15 June 2016).

13. Weidema, B.P.; Finnveden, G.; Stewart, M. Impacts from Resource Use—A common position paper. *Int. J. Life Cycle Assess.* **2005**, *10*, 382. [CrossRef]

14. Wenzel, H.; Hauschild, M.Z.; Alting, L. Methodology, Tools, Techniques and Case Studies. In *Environmental Assessment of Products*; Chapman & Hall: United Kingdom, UK; Kluwer Academic Publishers: Hingham, MA, USA, 1997; Volume 1, p. 544.

15. Guinée, J.B.; Gorrée, M.; Heijungs, R.; Huppes, G.; Kleijn, R.; de Koning, A.; van Oers, L.; Sleeswijk, A.W.; Suh, S.; Udo de Haes, H.A.; et al. *Handbook on Life Cycle Assessment: Operational Guide to the ISO Standards*; Kluwer Academic Publishers: Dordrecht, The Netherlands, 2002.

16. Schneider, L.; Berger, M.; Finkbeiner, M. Abiotic resource depletion in LCA—Background and update of the anthropogenic stock extended abiotic depletion potential (AADP) model. *Int. J. Life Cycle Assess.* **2015**, *20*, 709–721. [CrossRef]

17. Szargut, J. International progress in second law analysis. *Energy* **1979**, *5*, 709–718. [CrossRef]

18. Finnveden, G.; Östlund, P. Exergies of Natural Resources in Life Cycle Assessment and Other Applications. *Energy* **1997**, *22*, 923–931. [CrossRef]

19. Bösch, M.E.; Hellweg, S.; Huijbregts, M.; Frischknecht, R. Applying cumulative exergy demand (CExD) indicators to the ecoinvent database. *Int. J. Life Cycle Assess.* **2007**, *12*, 181–190. [CrossRef]

20. Dewulf, J.; Bösch, M.E.; De Meester, B.; Van der Vorst, G.; Van Langenhove, H.; Hellweg, S.; Huijbregts, M.A.J. Cumulative Exergy Extraction from the Natural Environment (CEENE): A comprehensive Life Cycle Impact Assessment method for resource accounting. *Environ. Sci. Technol.* **2007**, *41*, 8477–8483. [CrossRef] [PubMed]

21. Ayres, R.U.; Ayres, L.W.; Martinas, K. Exergy, waste accounting and life-cycle analysis. *Energy* **1998**, *23*, 355–363. [CrossRef]

22. Gössling-Reisemann, S. Combining LCA with thermodynamics. In *Information Technologies in Environmental Engineering*; Marx Gomez, J., Sonnenschein, M., Müller, M., Welsch, H., Rautenstrauch, C., Eds.; Springer: Heidelberg, Germany, 2007; pp. 387–396.

23. Valero, A. From grave to cradle. A thermodynamic approach for accounting for abiotic resource depletion. *J. Ind. Ecol.* **2013**, *17*, 43–52. [CrossRef]

24. Frischknecht, R.; Wyss, F.; Büsser Knöpfel, S.; Lützkendorf, T.; Balouktsi, M. Cumulative energy demand in LCA: the energy harvested approach. *Int. J. Life Cycle Assess.* **2015**, *20*, 957–969. [CrossRef]

25. Frischknecht, R.; Jungbluth, N.; Althaus, H.-J.; Doka, G.; Dones, R.; Hischier, R.; Hellweg, S.; Nemecek, T.; Rebitzer, G.; Spielman, M. *Overview and Methodology*; Final report ecoinvent data v 2.0, No. 1; Swiss Centre for Life Cycle Inventories: Dübendorf, Switzerland, 2007.

26. Classen, M.; Althaus, H.-J.; Blaser, S.; Scharnhorst, W.; Jungbluth, N.; Tuchschmid, M.; Faist Emmenegger, M. *Life Cycle Inventories of Metals*; Final report ecoinvent data v2.1; Swiss Centre for LCI, Empa—TSL: Dübendorf, Switzerland, 2009; Volume 10.

27. Frischknecht, R.; Tuchschmid, M.; Faist Emmenegger, M.; Bauer, C.; Dones, R. *Strommix und Stromnetz Sachbilanzen von Energiesystemen*; Final report No. 6 ecoinvent data v2.0; Dones, R., Ed.; Swiss Centre for LCI, PSI: Dübendorf/Villigen, Switzerland, 2007; Volume 6.

28. Hischier, R.; Classen, M.; Lehmann, M.; Scharnhorst, W. *Life Cycle Inventories of Electric and Electronic Equipment—Production, Use &amp*; Final report ecoinvent Data v2.0; Swiss Centre for LCI, Empa—TSL: Duebendorf/St. Gallen, Switzerland, 2007; Volume 18.

29. Szargut, J.; Morris, D.R.; Steward, F.R. *Exergy Analysis of Thermal, Chemical and Metallurgical Processes*; Hemisphere: New York, NY, USA, 1988.

30. De Meester, B.; Dewulf, J.; Janssens, A.; van Langenhove, H. An improved calculation of the exergy of natural resources for exergetic life cycle assessment (ELCA). *Environ. Sci. Technol.* **2006**, *40*, 6844–6851. [CrossRef] [PubMed]

31. Finnveden, G. On the Limitations of Life Cycle Assessment and Environmental Systems Analysis Tools in General. *Int. J. Life Cycle Assess.* **2000**, *5*, 229–238. [CrossRef]

32. Hall, C.A.S.; Cleveland, C.J.; Kaufman, R. *Energy and Resource Quality: The Ecology of the Economic Process*; Wiley-Interscience: New York, NY, USA, 1986.

33. Gössling-Reisemann, S. What is resource consumption and how can it be measured? *J. Ind. Ecol.* **2008**, *12*, 10–25. [CrossRef]
34. Sonnemann, G.; Gemuchu, E.D.; Adibi, N.; De Bruille, V.; Bulle, C. From a critical review to a conceptual framework for integrating the criticality of resources into Life Cycle Sustainability Assessment. *J. Clean. Product.* **2015**, *94*, 20–34. [CrossRef]
35. Isacs, L.; Finnveden, G.; Håkansson, C.; Steen, B.; Tekie, H.; Rydberg, T.; Widerberg, A.; Wikström, A. Valuation of abiotic resources in impact assessment. *Res. Conserv. Recycl.* **2015**. Submitted.
36. Achachlouei, M.A.; Moberg, Å.; Hochschorner, E. Life Cycle Assessment of a magazine, part I: Tablet edition in emerging and mature states. *J. Ind. Ecol.* **2015**, *19*, 575–589. [CrossRef]
37. Alanya, S.; Dewulf, J.; Duran, M. Comparison of overall resource consumption of biosolids management system processes using exergetic life cycle assessment. *Environ. Sci. Technol.* **2015**, *49*, 9996–10006. [CrossRef] [PubMed]
38. Reuter, M.A.; von Schalk, A.; Gediga, J. Simulation-based design for resource efficiency of metal production and recycling systems: Cases—Copper production and recycling, e-waste (LED lamps) and nickel pig iron. *Int. J. Life Cycle Assess.* **2015**, *20*, 671–693. [CrossRef]
39. Jeswani, H.K.; Azapagic, A.; Schepelmann, P.; Ritthof, M. Options for broadening and deepening the LCA approaches. *J. Clean. Product.* **2010**, *18*, 120–127. [CrossRef]
40. Portha, J.-F.; Louret, S.; Pons, M.-N.; Jaubert, J.-N. Estimation of the environmental impact of a petrochemical process using coupled LCA and exergy analysis. *Resour. Conserv. Recycl.* **2010**, *54*, 291–298. [CrossRef]
41. Hiraki, T.; Akiyama, T. Exergetic life cycle assessment of new waste aluminium treatment system with co-production of pressurized hydrogen and aluminium hydroxide. *Int. J. Hydrog. Energy* **2009**, *34*, 153–161. [CrossRef]
42. Yang, L.; Zmeureanu, R.; Rivard, H. Comparison of environmental impacts of two residential heating systems. *Build. Environ.* **2008**, *43*, 1072–1081. [CrossRef]
43. De Meester, B.; Dewulf, J.; Verbeke, S.; Janssens, A.; van Langenhove, H. Exergetic life-cycle assessment (ELCA) for resource consumption evaluation in the built environment. *Build. Environ.* **2009**, *44*, 11–17. [CrossRef]
44. Wall, G. Exergy conversion of the Swedish society. *Resour. Energy* **1987**, *9*, 55–73. [CrossRef]
45. Chen, G.Q.; Qi, Z.H. Systems account of societal exergy utlilization: China 2003. *Ecol. Model.* **2007**, *208*, 102–118. [CrossRef]
46. Dewulf, J.; van Langenhove, H.; Muys, B.; Bruers, S.; Bakshi, B.R.; Grubb, G.F.; Paulus, D.M.; Sciubba, E. Exergy: Its potential and limitations in environmental science and technology. *Environ. Sci. Technol.* **2008**, *42*, 2221–2232. [CrossRef] [PubMed]
47. Liao, W.; Heijungs, R.; Huppes, G. Thermodynamic analysis of human-environment systems: A review focused on industrial ecology. *Ecol. Model.* **2012**, *228*, 76–88. [CrossRef]
48. Finnveden, G.; Nilsson, M.; Johansson, J.; Persson, Å.; Moberg, Å.; Carlsson, T. Strategic Environmental Assessment Methodologies—Applications within the energy sector. *Environ. Impact Assess. Rev.* **2003**, *23*, 91–123. [CrossRef]
49. Szargut, J.; Morris, D.R. Cumulative exergy consumption and cumulative degree of perfection of chemical processes. *Int. J. Energy Res.* **1987**, *11*, 245–261. [CrossRef]

resources

MDPI

Article

Dematerialization—A Disputable Strategy for Resource Conservation Put under Scrutiny

Felix Müller [1,*], Jan Kosmol [1], Hermann Keßler [1], Michael Angrick [2] and Bettina Rechenberg [3]

[1]   Section for Resource Conservation, Material Cycles, Minerals and Metal Industry, German Environment Agency, Wörlitzer Platz 1, 06844 Dessau-Roßlau, Germany; jan.kosmol@uba.de (J.K.); hermann.kessler@uba.de (H.K.)

[2]   Division for Emissions Trading, German Emission Allowance Trading Authority, German Environment Agency, Bismarckplatz 1, 14193 Berlin, Germany; michael.angrick@uba.de

[3]   Division for Sustainable Production and Products, Waste Management, German Environment Agency, Wörlitzer Platz 1, 06844 Dessau-Roßlau, Germany; bettina.rechenberg@uba.de

*   Correspondence: felix.mueller@uba.de; Tel.: +49-340-2103-3854

Received: 5 October 2017; Accepted: 24 November 2017; Published: 4 December 2017

Abstract: Dematerialization is a paradigm in resource conservation strategies. Material use should be reduced so that resource consumption as a whole can be lowered. The benefit for humankind should be completely decoupled from the natural expenditure by a definite factor X. Instinctively, this approach is convincing, because our entire value-added chain is based on material transformation. Targets for mass-based indicators are found within the context of justification for ecological carrying capacity and intergenerational fairness, taking into account the economic and socio-political expectation of raw material scarcity. However, in light of further development of material flow indicators and the related dematerialization targets, the question arises as to what they actually stand for and what significance they have for resource conservation. Can it be assumed that pressure on the environment will decline steadily if the use of materials is reduced, whether for an economy or at the level of individual products or processes? The present narrative review paper has discussed this issue and takes into account the authors' experience of the extended political and scientific discourse on dematerialization in Germany and Europe. As a result, a high "resource relevance" cannot be inferred from high physical material inputs at any of the levels considered. It has been shown that establishing mass-based indicators as control and target variables is questionable and that dematerialization exclusively based on such indicators without mapping other resources should be critically examined.

Keywords: natural resources; mass-based indicators; dematerialization; MFA; raw materials; resource conservation; resource efficiency; criticality; area of protection; precautionary principle

1. Introduction

Biophysical methods and models suggest that a global safe operating space can be identified where safe and sustainable life is granted to humankind without exceeding certain ecological limits [1]. However, a degradation of ecosystems and ecosystem services shows that those ecological limits have already been reached [2]. Moreover, humankind has most probably already exceeded the stress limits for climate change, phosphate and nitrogen nutrient fluxes, land-use changes, and biosphere integrity—four out of nine specified planetary limits [3]. If this situation is not resolved effectively, it must be assumed that not only will the Earth's state gradually worsen but also change irreversibly with devastating consequences for life on this planet.

Making express references to the planetary boundaries as well as resource scarcity and availability, national and international political agendas are increasingly migrating towards acceptance of resource conservation with a prime focus on increasing the efficiency of its use. European activities—starting from the milestone of the "Thematic Strategy on the Sustainable Use of Natural Resources" in 2005—have provided an implementation roadmap for the "resource-efficient Europe" flagship initiative [4–6]. This initiative is aimed at simultaneously enhancing economic performance and reducing the associated resource consumption. The European Environment Agency (EEA) has published a comprehensive report to show how 32 European states want to increase their resource efficiency [7]. The Federal Government of Germany stipulated its own resource efficiency programme *ProgRess* in 2012 [8]. An evaluation of performance was published on the basis of a progress report in 2016, and a second resource efficiency programme was adopted at the same time [9]. Countries such as Japan developed national strategies [10] many years ago, or—as in the case of India [11]—have recently developed one.

At an international level, in founding the *Resource Efficiency Alliance*, the G7 countries have committed to ambitious measures to protect natural resources and improve resource efficiency, building essentially on the reduce–reuse–recycle framework of the *Kobe 3R Action Plan* and existing national initiatives [12]. Even more importantly, the G20 launched a *Resource Efficiency Dialogue*, which will exchange good practices and national experiences to improve the efficiency and sustainability of natural resource use across the entire life cycle and to promote sustainable consumption and production patterns [13]. Resource efficiency is most prominently represented by the UN Sustainable Development Goals (SDGs). The 2030 Agenda for Sustainable Development envisages a world in which consumption and production patterns and use of all natural resources are sustainable [14].

In recent years, the concept of a circular economy has received increasing attention worldwide due to the recognition that the security of supply of resources and resource efficiency are crucial for the prosperity of economies [15]. In the meantime, existing strategies for resource efficiency have already been supplemented or even replaced by those for a circular economy, although there is a very large congruence in terms of contents. The EU Commission has presented a new circular economy strategy in line with the existing 2020 Europe Strategy flagship initiatives [16]. The Strategy consists of an EU Action Plan for the Circular Economy, including revised legislative proposals on waste that establishes a concrete and ambitious programme of action, with measures covering the whole cycle: from production and consumption to waste management and the market for secondary raw materials. A declared goal is to implement the paradigm shift from the resource-intensive, linear, "take, make, consume, and dispose" economic model to a resource-efficient circular economy. Initiatives, strategies, programmes, and roadmaps are being pursued to help further the development of a circular economy by organizations, and in many countries such as Finland [17] and the Netherlands [18], with the aim of strengthening a resource-efficient economy [19]. Even if the distinction between resource efficiency and circular economy is not clear, circular economy can be described as a desirable target state of a resource-conserving economy.

In the programmes and strategies mentioned, there is as yet no generally accepted definition of the central object, the natural resources. Depending on their definition and interpretation, they include materials, raw materials, natural energy flows, land, wildlife, agricultural goods, environmental media, ecosystems, the services they provide, and biodiversity. This results in more or less major overlaps between the natural environment, natural resources, natural capital, and, as a consequence, resource use and environmental impact. The strategies often use the terms of resource use and resource consumption synonymously or arbitrarily. This unclear terminology is not solely linguistic but raises fundamental conceptual questions about the relationship between environmental protection and resource conservation and the target parameters of resource conservation.

It is striking that, although a "broad understanding" of natural resources is often considered, an implicit or partly explicit focus is placed on raw materials or materials. For example, those indicators concerned with the frugal use of fresh water, land areas, or the protection of genetic diversity are also included in the final list of coordinated SDG indicators. However, the material footprint or its productivity was chosen as the only indicator to represent natural resources under the central SDG 12 meant to achieve sustainable management and efficient use of natural resources by 2030 [20]. Similar material flow indicators and their productivities have also been described in national programmes and sustainability strategies, or even prominently set as leading indicators for the programmes to measure the success of resource conservation or resource efficiency [21–24].

The widespread focus on material in strategies for natural resource use goes back to a 1990s school of thought that considers dematerialization, i.e., a reduction of material use by a "factor X", as a guiding principle for a sustainable economy. Such an economy avoids negative environmental impacts and faces scarcities in a precautionary way, in responsibility as well, for future generations. The declared objective is to decouple benefits from resource demand, which is equated to material demand in this school of thought. This dematerialization approach is propagated not only at the overall economic level but also at the micro level for products and processes [25]. The discussion of approaches to dematerialization has given a great boost to the development of material flow calculations and accounting [26,27]. However, the significance of the concept and its material-based indicators is considered controversial [28–32]. The objectives are the subject of political, societal, and scientific discussions. Although, to the authors' knowledge, there are so far no binding resource policies that explicitly aim at dematerialization, the above-mentioned focus on material efficiency and the common synonymous use of the terms resources and material makes an implicit reference to this school of thought.

Research Questions and Goals

In light of further developments of the macroeconomic material flow indicators for resource efficiency assessment and of the analysis of material flows at process and product levels, questions arise as to what mass-based indicators stand for and what they reveal about the conservation or the efficient use of natural resources. Instinctively and emotionally, they are highly relevant. Our entire value-added chain is based on material and substance transformation: we extract material from the environment, transform it, and permanently discharge emissions and waste into the natural environment, which influences its state. Even services and non-material goods in the strict sense such as education are dependent on these material transformation processes: they require infrastructure such as roads, railways, heated buildings, computer centers, and much more. In turn, these can be used to fulfil their purpose only with auxiliary materials and fuels, mostly fossil fuels.

If so much depends on our material flows, does reduction of material use by a factor X suit as a guiding principle for a sustainable economy? Is it safe to assume that environmental pressure will drop steadily and resource scarcity be addressed if material use is reduced across the board by a certain factor X, as the dematerialization school of thought is assuming? How reliable is the measurement of the success of resource conservation using dematerialization indicators in kilograms, tonnes, and megatonnes and can any connection be made to the relevant areas of protection? Regarding the breadth of all natural resources, how "resource-relevant" are mass-based indicators? Do they qualify as control variables for policies that aim at conservation or the efficient use of resources?

Against this background, this paper presents a narrative review of relevant methodologies and assessments for raw materials. It discusses the current application of material flow indicators and juxtaposes concepts for raw materials and safeguard subjects from a micro level. It reflects the shortcomings, advantages and main criticism with regard to methodology and transported messages. Thereof, conclusions are drawn about the use and interpretation of mass-based indicators for resource policy.

The structure of this paper is as follows: Section 2 discusses the technical background. The authors present their classification of resource conservation terminology. The application of economy-wide material flow accounting (EW-MFA) indicators in political programmes is described using Germany as an example. This is followed by an outline of the factor X approach and its fundamentals for dematerialization as a strategy. In Section 3, the relevance of material flow indicators for the use of natural resources is discussed on the basis of the author's experience. Areas of protection and impact paths of raw material consumption are described. The authors discuss the relevance of material flow indicators for environmental impacts and raw material scarcities, which are at the core of the justification for dematerialization and resource efficiency strategies. The extent to which it is legitimate to use the precautionary principle to justify dematerialization strategies and to derive dematerialization targets is additionally discussed. In Section 4, the key findings are summarized, followed by a concluding discussion to put the results into perspective in the resource policy debate in Section 5.

2. Resource Conservation Conceptualization for Materials and the Factor X Approach

2.1. Terminology

The resource conservation policy field has so far been strongly anthropocentric and utilitarian, which is already laid out in the semantics of the resource term and is reflected in different definitions of the term natural resources [33–37] and in explanations on the subject [6,8,38–40]. This means that this policy field does not focus on nature's intrinsic value and its living beings or the needs of other sentient creatures, but rather on the needs of humans or benefit or value for humans. Natural resources create the material, energy and spatial basis for human life and are indispensable for our economy and welfare. We satisfy our needs by using them as sources of energy, production means, living and recreational areas, a pool of pharmaceutical active agents, and so on [40]. All these natural resources contribute to our natural capital, which must be conserved and enlarged over the long term.

Other definitions do not make this explicit reference to human needs or benefit but define them as "all components of nature" [9] or are confined to the final and non-exhaustive list of their components [4,9,41,42]. Depending on the definition, natural resources include at least the geological stocks of primary raw materials, defined as unprocessed materials or material mixtures which are extracted from nature to be input into production processes [43]. In addition, physical space including area, environmental media such as water, soil, and air, flow resources such as solar radiation, wind and water flows, ecosystems, and ecosystem services for humans are also counted as natural resources [6].

One of the few recent definitions that are not exhausted in an enumeration of its components defines natural resources as "means found in nature that can be beneficial for humans" [43]. This paper uses this definition. This conception of natural resources includes environmental media and ecosystems, not only in their function as source of material and energy, but also as a sink for emissions (sink function) and other functions. They bring us services on a daily basis, often invisibly—be it the soil's water filter function, the rivers' self-cleaning capacity, the atmosphere's emission absorption capability, flood control by floodplain forests, the regeneration capability of fish stocks, or the pollination by wind, water, and animals—and must therefore be considered natural resources in the sense that they are "means found in nature that can be beneficial for humans".

From this broad understanding of natural resources, however, delineation questions arise about the natural environment and, subsequently, conceptual issues about the relationship between environmental protection and resource conservation. Emissions into environmental media and ecosystems are, in this sense, not only environmental impacts but also a resource use. According to this understanding, resource conservation is merely another, strictly utilitarian–anthropocentric perspective of environmental protection. This is followed by a further terminological blurring of resource conservation, which is of decisive relevance for the conceptual development of this field of science and policy: Is resource conservation aiming at reducing resource use or resource consumption? Resource use means any human access to natural resources, while resource consumption is a form of resource use that converts resources in such a way that they are not available for any other uses [43].

While resource conservation designates the careful handling of natural resources in order to avoid their depletion or consumption, resource efficiency or resource productivity is the ratio between a given benefit or result and the natural resource use required to achieve it [43–45]. Resource efficiency is conceptually based on the decoupling concept shown in Figure 1. However, the arguments in the last paragraph raise the question whether or not a double decoupling concept is reasonable and necessary, aiming at decoupling both prosperity and human well-being from resource use as well as resource use from environmental impacts (see Figure 1). In contrast to the resources, the benefit is a rather subjective or judgemental category. "The benefit of products and services is given by functions that are frequently described by technical parameters. In economy, the benefit is frequently expressed in monetary terms and given as turnover or product price numbers. Both the technical and monetary assessments of benefit do not fully cover all forms of all benefit aspects" [43]. Since the political resource efficiency debate aims to address economic performance in relation to natural resource use (see Section 1), this article uses the term benefit, which includes technical function but allows monetary valuation.

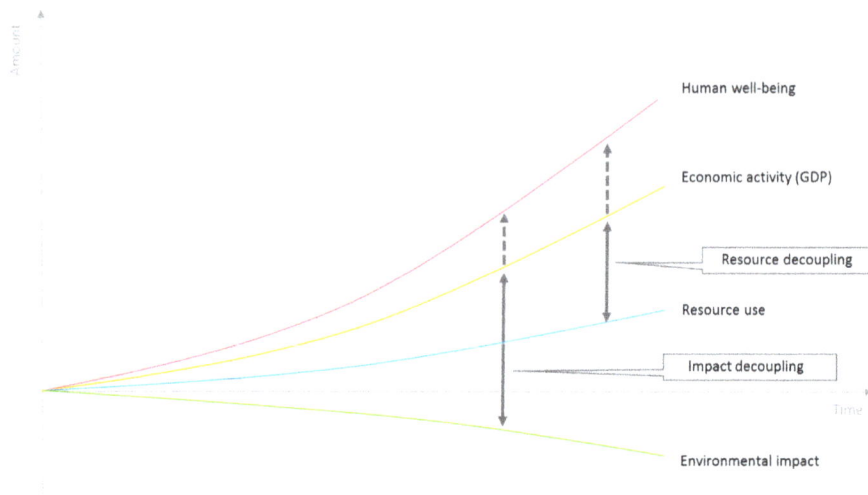

Figure 1. Absolute and realtive decoupling pathways for well-being, economic activity, resource use, and environmental impacts [46]. In the most desirable case, benefits, i.e., GDP, increase while cost, i.e., environmental impact or resource use, decreases. This constellation is called absolute decoupling. Even if cost in terms of resource use increases, productivity can still increase if benefits increase more strongly. This constellation is also called relative decoupling. A third case is also conceivable: if both cost and benefit decrease, but cost decreases more than benefit, productivity increases.

2.2. Productivity Indicators and Decoupling by Germany's Example

Raw material productivity, a key indicator of the German sustainability strategy from 2002, is a basic element of the German resource efficiency programme [8,9]. Raw material productivity sets the gross domestic product in relation to the abiotic part of direct material input (DMI_{abiot}) from domestic raw material extraction and imports. DMI is an input indicator in the system of economy-wide material flow indicators (Figure 2). Input indicators map the (raw) material inputs used to provide all services of an economy. If imports are recorded not only in their own weight as in the case of the DMI but also together with the used or even unused foreign extraction, the indicators Raw Material Input (RMI) or Total Material Requirement (TMR) are used for this purpose. The same metabolic system also specifies indicators that map domestic consumption using the DMC, RMC, or TMC indicators, and physical trade balance or stock growth by balancing inputs and outputs in different ways [47].

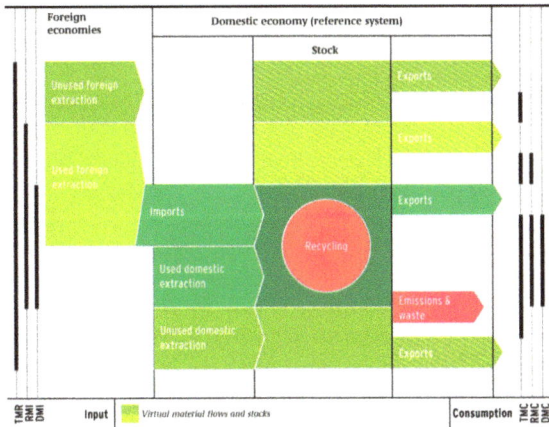

Figure 2. Illustration of economy-wide material flow indicators. The lateral lines indicate the respective system boundaries and balancing items [40].

The indicator for raw material productivity provides information on the efficient handling of abiotic materials. Through its relationship to GDP the indicator is chiefly considered to be an economic parameter for the production factor of abiotic materials. Raw material productivity is thus analogous to labor and capital productivity.

GDP and DMI_{abiot} show an absolute decoupling over time (Figure 3). However, DMI_{abiot} takes into account imports at their actual processing level, the so-called direct material flows [47]. Those raw materials that were used for the production of imported goods in other countries beyond the dead weight of imports are not accounted for. If resource-intensive processes are relocated abroad and highly processed goods are imported instead, raw material productivity increases [48]. Since the tendency is that semi-finished and finished products are increasingly imported, the observed absolute decoupling is mainly due to the indicator's methodological weakness [49].

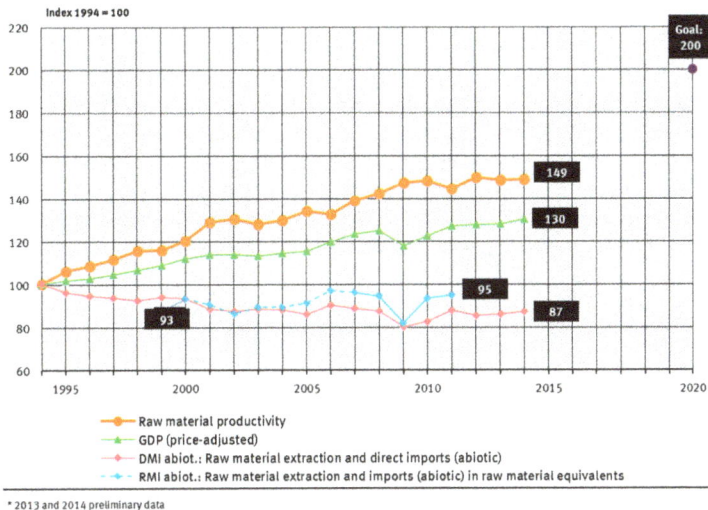

Figure 3. Raw material productivity in Germany from 1994 to 2014 and the underlying trends of GDP, DMI_{abiot}, and RMI_{abiot}. Data based on [50].

Since 2007, a methodology has been developed to cover raw materials used for imports [49,51,52]. This methodology uses input–output analyses and process chain approaches to identify goods together with their raw material equivalents (RMEs). On average, the RME is 2.7 tonnes per 1 tonne of directly imported material into Germany, which means that 2.7 tonnes of raw material are needed abroad to produce 1 tonne of physically imported goods [48]. As a comparison, the corresponding imports in RME for the EU-28 on average are estimated as being 2.4 times higher than actual physical imports, thus in the same order of magnitude as in Germany [53]. The primary raw material input (RMI) indicator maps the sum of domestic raw material extraction and direct and indirect imports beyond the dead weight of the physical imports (Figure 2). RMI_{abiot} is only relatively decoupled from GDP over time [21]—in contrast to DMI_{abiot} (Figure 3). If consumption indicators are taken into account, physical material flow indicators such as domestic material consumption (DMC) may even exhibit diametrically opposed trends to their analogous indicators in RME (raw material consumption, RMC), i.e., a decrease versus a rise [22].

2.3. "Factor X"

In the context of resource conservation policy, firm objectives for dematerialization are being discussed. These are linked to different material flow indicators (Figure 2) and have very different ambition levels. Bringezu and Mancini et al. provide an overview of more recent proposals [30,54]. The dematerialization objectives are rooted in the tradition of a so-called factor X concept. X stands for a reduction factor in the use of a resource for providing a service. Both should be decoupled from each other. This corresponds to the concepts and logic of eco- and resource efficiency or respective productivity [31]. The factor X approach is applied not only to national economies but also to products, services, sectors, industries, and needs [55]. Precursors of the factor X approach can be found in Daly [56] and Ayres [57] whose environmental models also emphasized the necessity of dematerialization, but above all a significant increase in productivity of natural resources. The actual pioneering work of the factor X approach includes a report to the Club of Rome by von Weizsäcker et al.: "Doubling wealth—halving resource use" [58] and Schmidt-Bleek's book: "How much environment does man need?" [59]. While the two Wuppertal Institute scientists estimated necessary enhancement factors "X" to 4, 5 [60], and 10, other factors of up to 50 have been postulated [31]. The factor targets can hardly be compared since they have different base years depending on the publication date and they extend to very different time periods, from a few decades up to 100 years. Moreover, the authors do not often believe that target horizons and sizes can be made more precise but rather aim at target corridors and orders of magnitude.

Very different perceptions are behind the factor approaches as to which kind of resource use should be reduced. Schmidt-Bleek linked a factor 10 to the concept of material intensity per service unit (MIPS) that he developed, i.e., the reciprocal of productivity. Here, abiotic and biotic materials, including air, and water, are balanced over the service life cycle of products and services. This concept explicitly includes the so called "unused extraction" such as overburden in mining, soil movements in agriculture and forestry, erosion as well as soil excavation for construction [61–63]. The categories mentioned minus air and water are also called primary materials and are mapped by the material flow indicator total material requirement (TMR) at the national economic level (see Figure 2). The Factor 10 concept has also been promoted by the identically named *Factor 10 club*—starting with a Carnoules Declaration in 1994 [64]. However, the references are too general and ambiguous in a terminological sense. This leads to an interpretation mismatch as to which resource use or input is being addressed (see Section 2.1).

Schmidt-Bleek's MIPS-concept is at the core of dematerialization as a strategy for resource conservation [25,59,61,65]. It is based on the assumptions that the planetary ecological carrying limits have been exceeded, each man-made material flow causes changes in the ecosphere, and the volume of man-made material flows (extraction of material from the ecosphere) exceeds the extent of natural material flows at the earth's crust (volcanism, erosion, etc.). Furthermore, Schmidt-Bleek bases the necessity for dematerialization on the realization that man is a long way from fully understanding the cause–effect relationships in the interaction between man and nature, our knowledge about specific environmental pressures is extremely limited, and environmental problems are only recognized as such after their occurrence.

Therefore, Schmidt-Bleek proposed that, in addition to emissions and wastes, the focus of environmental policy must be directed towards material inputs into the industrial metabolism since the laws of conservation of mass and energy imply that any input (material extraction) would inevitably become an output (waste, emission, etc.) and thus would address the total environmental impact potential of human activities, including the still unknown environmental problems. This hypothesis has been questioned by various authors [29,66–68]. Schmidt-Bleek deduces from the precautionary principle the premise that natural systems should be changed by human activities as little and as slowly as possible and justifies this with the above-mentioned assumptions about the connection between material input and environmental impacts and the wide ignorance of cause–effect relationships. He states that, by lowering the industrial metabolism to "an ecologically compatible degree", not only known and unknown environmental pressures due to raw material extraction and processing but also environmental impacts during production, use, and disposal of the products produced therefrom should be reduced or avoided, and an "evolutionary balance", "ecological balance", or "ecological stability" reached [59].

Other factor approaches such as those of von Weizsäcker do not make an explicit reference to material flow indicators and are not restricted to rigid dematerialization as his approach focuses on eco-efficiency [58]. The action proposals are backed by great technological optimism. Energy demand and greenhouse gas emissions have been chosen as the design basis. However, it may be considered a problem that a factor target for certain resources can be achieved, while it may clearly be missed or even opposing effects may arise for other natural resources within the same system.

Factor approaches must be scrutinized in certain cases because contrasting ambition levels with different societal, economic, and political implications may result from numerically identical factors depending on the consideration's system boundaries, resources addressed, indicators applied including their specified balance limits, the periods considered, the stipulation of absolute targets or productivities, which may also be substantially driven by autonomous GDP developments (see Section 2.2).

The discussion on dematerialization initiated by Schmidt-Bleek and others in the 1990s has given a great boost to the development of material flow calculations and accounting [26,27]. Mass-based indicators and their productivities are at the heart of most resource-efficiency policies (see Section 1) and make an implicit reference to Schmidt-Bleek's hypothesis that aggregated mass would constitute a reasonable proxy for potential environmental impact. According to the perception of the authors, this bold interpretation of the significance of mass-based indicators still endures in debates on resource efficiency and conservation and is an obstacle for successful resource policies.

3. Assessment

3.1. Areas of Protection

The area of protection (AoP) that is addressed by primary raw materials is ambiguous. What exactly must be prevented by using less primary raw materials [69]? Can this also be best represented by means of mass-based indicators? An insight into the widely developed method discussion of life cycle assessment since the beginning of the 1990s helps answer these questions [70]. The logic of life cycle assessment is about the relative reduction of environmental pressures by individual products or processes. However, in order to cover the impact on the environment as a whole, it is described using AoP. These AoPs are damaged by relevant environmental pressures [71]. However, life cycle assessment does not specify which AoP must be respected or why [72]. This issue is addressed by a school of thought that logically argues that the environment is worthy of protection and must first be defined top-down according to the value system in society before its influence and damage can be assessed.

In the early phases of life cycle assessment, the Society of Environmental Toxicology and Chemistry (SETAC) recognized that raw materials are part of the environment worthy of protection [73]. Recently however, different AoP concepts have still led to very diverse approaches for interpretation and definition. The evaluation and preferred selection of certain methods is the subject of on-going discussions [74]. The fundamental anthropocentric context of justification for the raw materials consumption as part of the AoP natural resources can be seen less in the ecological impacts—represented by the ecosystem quality—or human health—which is an AoP on its own—but more in the actual consumption, availability, and possible shortages of raw materials [75,76]. The interest in AoPs can go beyond the purely ecological sector [77]. Societal and social welfare or the fundamental adherence to the precautionary principle is also mentioned in this context [75,78]. Both the definition and the logical connection between AoPs contain normative stipulations that stem from anthropocentric, pathocentric, and biocentric positions [79]. Such issues of value judgement can also be supported by ethical, theological, legal, and economic arguments [71]. Many established ecological assessment methods allocate environmental impacts to the three AoPs: human health, ecosystem quality, and resource consumption [80]. Some authors argue that the man-made—anthropogenic—environment should also be addressed as an independent AoP that inherently goes beyond the ecological sphere [81]. In that sense, Dewulf proposes five different perspectives of natural resources, which comprise not only the natural resources' asset, their provisioning capacity, and their role as constituents in global ecosystem functions but also their provisioning function for human welfare and human welfare as an inclusive category itself [78].

The AoPs are intrinsically linked to one another. Thus, the extraction and processing of raw materials inevitably leads to reactions with the state of the ecosystem and to a threat to health, for example, through soil changes and mobilized heavy metals. An acidifying emission can cause damage to buildings and thus damage the physical capital of the anthropogenic environment and impair human welfare (see Section 3.2.1). These relationships lead to the fact that resource use and the associated material changes and shifts can subsequently affect different safeguard subjects within the AoP.

There are four basic notions for the assessment of raw materials use and their interpretation as safeguard subjects in the AoP natural resources that are of particular interest [71].

1. **Intrinsic values**: Infringing the supply of raw materials as such is understood as an inadmissible intervention. Each raw material extraction is therefore to be regarded as detrimental to the protection objective.
2. **Reduction**: Reducing the raw materials' availability is regarded as a restriction on the freedom of future generations. In this case, the decision-making freedom of future generations and intergenerational fairness are the overriding protection objective.
3. **Devaluation**: It is assumed that raw material deposits of lower quality must be exploited in future. This will cause increased mining costs for future generations.
4. **Substitution**: From this perspective, the demands of future generations will be equated with today's consumption interests and the substitution options of non-renewable raw materials will be regarded based on the technical and economic costs for developing alternative technologies.

Although denoted differently, these are congruent to the categories of impact pathways for resource depletion for raw material use, indicator approaches and addressed safeguard subjects in the AoP natural resources [82]. Accordingly, the scope of ideas ranges from primary raw materials as safeguard subjects over the purely physical availability for future generations and the anticipated impairment by the decreasing quality of deposits and resources, to the functional, intergenerational fulfilment of purpose as an actual safeguard subject from an anthropocentric consumer perspective.

Indicators in the first category are comparatively easy to determine at the life cycle inventory level. However, these indicators do not reveal the scarcity of raw materials, their thermodynamic properties, the time-related changes to deposit qualities, or the functional and monetary utility value of the extracted raw materials. This considerably limits their relevance to the safeguard subject of raw material consumption [82]. In contrast, the depreciation of raw material supply and the substitution of raw materials are of particular interest since they both try to measure actual damage as end points. The rationales clearly go beyond environmental considerations [78] into life cycle sustainability assessment (LCSA) [74,75].

Economy-wide material flow indicators such as RMI or RMC (see Section 2.2, Figure 2) and the methodically equivalent cumulative raw material demand (CRD) that collect the sum of the primary raw materials used to produce and transport a product along the value chain at the level of products and processes can clearly be assigned to a natural resource: primary raw materials. These indicators enable all material systems to be traced back to the basic extraction and exploitation of primary raw materials from the natural environment. The indicators explicitly address the AoP natural resources in terms of raw material consumption for biotic and abiotic raw materials minus water and air in units of mass without further characterization. Since no impact characterization or weighting that may project a scarcity information has been carried out, they can be arranged unequivocally into the context of justification: an intrinsic value is attributed to raw materials, expressed by its mass, and any form of raw material extraction and removal ex aequo impairs the AoP. Resources and deposits of all raw materials are thus regarded equally worthy to protect, irrespective of raw material availability. The indicators map this issue reliably and contradiction free and are appropriate to cover the mass of primary raw materials used. The change of state over several periods indicates whether reserves and resources (biotic extractions and crops) are actually more or less used. This is based on a rather egalitarian assessment approach, which assumes that it makes no difference at which time the raw material is extracted and that each removal can trigger the same potential environmental impacts in a damage model [83,84].

3.2. Relevance and Representativeness of Mass-Based Indicators

As mentioned in Section 1, mass-based material flow indicators are assigned a strong communication function in an information hierarchy as indices or key indicators to draw attention and possibly represent other natural resources [24]. The more these indicators are aggregated, the more they are subject to a target conflict between their information content and their effectiveness in communication [85]. This is illustrated by the information pyramid in Figure 4. By choosing and limiting the analysis to one or a few indicators, other key figures of a system that may also be important are disregarded. In the lower part of the pyramid, material flow indicators are analytically far more interesting for the analysis of the socio-economic metabolism. In contrast to the use of highly aggregated mass-based indicators the concept of the socio-economic metabolism also established as anthropogenic [86] or industrial metabolism [87,88] incorporates MFA-indicators in a systemic way and puts emphasis on the actual transformation of materials [26,27]. For complex systems such as national economies, material flow indicators, in their functional aggregation and classification according to different materials and uses, enable considerable insights into the system conditions, changes, and driving forces [89]. They provide basic structural information about the economy's material basis. Thus, input indicators linked to output indicators provide important systemic information and control variables for exports and inputs into the environment, stock accounting balances and domestic use, which can also be interpreted by the comparison of different economies and their development. The analytical instruments are essentially based on material flow analyses, accounting and modeling for comprehensive analysis.

However, when highly aggregated mass-based indicators are used at process and product levels purely solitarily or at a macro-economic level as headline indicators for natural resources, they are in the context of justification of economic necessities, ecological requirements, and intergenerational fairness and thus scarcity expectation is significant for economic and societal policy [90]. Beyond the intended signal effect and communication function, two crucial questions arise: How good are those mass-based indicators at representing environmental impacts and scarcities?

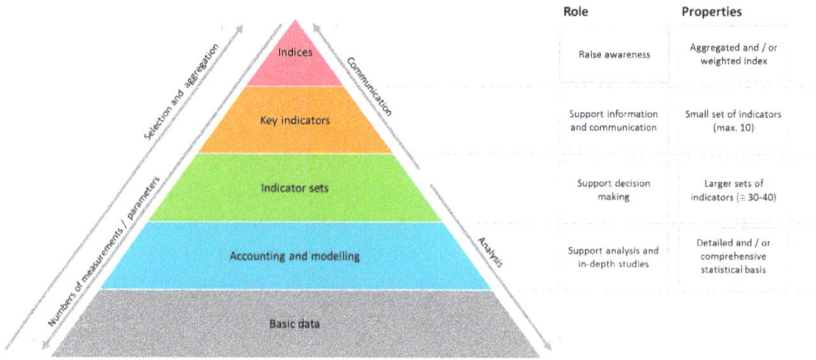

Figure 4. Information pyramid where indicators exercise multiple roles according to their position. Based on [45].

3.2.1. Representation Function for Environmental Impacts

In the environmental policy relevant DPSIR model (Driver-Pressure-State-Impact-Response), mass-based material flow indicators represent environmental pressures that can cause undesirable effects and condition changes (Figure 5). The DPSIR model is suitable for locating material flows in the human–environment set of interactions [91]. This creates an interdependency between socioeconomic drivers, environmental pressures and impacts, the state of the natural environment and the ecological effects or actual damage. Furthermore, it enables an integrated view of the societal, political, and economic measures that have an impact on the overall set of interactions.

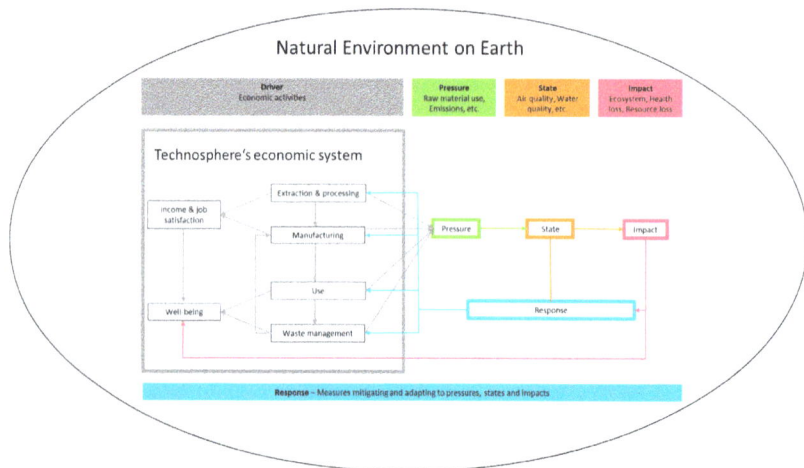

Figure 5. DPSIR-Framework that shows the interplay between the environment and socio-economic activities, based on [91,92].

Translated into the model logic of DPSIR, a dematerialization strategy now suggests that fewer raw materials are used when modified technologies or changed demand profiles in the economic system are employed; consequently, not only adverse, degrading changes in the environmental status but above all negative environmental impacts are avoided. As a result, fewer responses and adaptive measures would be required. The approach of dematerializing at all can also be interpreted as a response to an already experienced or expected over-use of natural resources, as suggested by Factor X approaches (see Section 2.3). If this is supposed to be promising, mass-based indicators would have to correlate with the actual environmental effects, in spite of the lack of effectivity characterization, i.e., they must be ecologically significant. The ecological significance of raw materials extends to the assessment of further use of natural resources through pressures on the environment (land use and conversion, energy demand, fresh water use, and consumption) and, in particular, environmental impacts (use of the environmental media sink function provided as an ecosystem service and impairments of biodiversity).

In the following, five scopes are to be systematically differentiated in order to correlate the raw material use of System S 1 with environmental impacts (Figure 6). Opposed to Scopes 1, 3, and 5, Scopes 2 and 4 are based on a consideration of the RME; i.e., cumulative raw material requirements up to the actual extraction and thus the physical pre-chains are taken into account. On the other hand, a kilogram of gold and a kilogram of steel would be included in equal measure in a correlation analysis in approaches 1, 3, and 5. However, the actual specific raw material requirement per ton of material (primary raw material intensity) as defined by the indicator CRD (see Section 3.1) of basic materials and homogeneous semi-finished and finished products increases in the following order, as an initial approximation: construction minerals < fuels < biomass < industrial minerals < basic chemicals and plastics < ferrous and non-ferrous metals < special metals < precious metals [93]. In detail, however, there is a special rank sequence for each environmental pressure or impact category (e.g., land use, energy use, and greenhouse gas emissions) at the impact level of the respective materials [94,95]. With regard to environmental pressures and impacts, Figure 6 shows Scopes 1 and 2 as well as Scopes 3 and 4 forming pairs. The crux is whether material or primary raw material use in System S 1 is only correlated to the impacts in the upstream chains (for example, by the extraction, refining, and casting of a semi-finished metal product outside System S 1, under consideration) or additionally with the effects which are described in the following material use in System S 2 (for example, by a car in which said semi-finished metal product has been installed). The dematerialization logic of the factor X concept, "any input would become an output", suggests the latter (see Section 2.3).

The narrow Scope 5 according to Figure 6 was pursued in a study for the EU 27 that was aimed at identifying threshold indicators in seven environmental action fields [96]. The extent to which direct abiotic domestic materials consumption (DMC_{abiot}) can serve as a proxy indicator for national emission limiting values for different air pollutants was investigated, and it was compared with the emissions reported according to the territorial principle of the NEC Directive. However, no linearity could be found which would indicate that DMC_{abiot} could represent achieving the national emission limiting values. Neither DMC_{abiot} nor emission data take account of processes abroad. Therefore, they methodically correspond to each other, although they allow only a limited perspective of the overall resource consumption.

Numerous studies have performed regression analyses according to Scopes 1 and 2 in Figure 6 to systematically check the representation function of mass-based and further resource indicators for other natural resources (e.g., representativeness of (raw) material input for land-take or the emission of greenhouse gases). In their study, based on 130 materials and products, Giegrich et al. [97] concluded that pure mass-based material flow indicators (RMI and TMR) are not suitable as representative resource indicators. A general representation function of individual indicators for the other resource indicators has not been established. This applies neither to other input resources (especially energy, land, and water) nor to the use of the sink function. Other studies that performed similar correlation analyses on up to 100 materials based on their life cycle inventories have arrived at comparable results [98,99]. In addition to raw materials, the physical input resources such as land, water, and energy must be dealt with separately in order to map the use of physical resources with sufficient reliability [97].

If a single environmental impact indicator is supposed to represent environmental impacts, the most appropriate indicator is the primary energy demand according to the results. The latter is also the result of a very complex regression analyses on up to 1200 products and materials. Huijbregts et al. [100,101] found that non-renewable energy demand represents many environmental impacts fairly well. Use of non-renewable fossil and nuclear fuels with unambiguous qualitative interactions can be regarded as a key driver for many well-known negative environmental impacts. Conclusions about an inadequate representation function of mass-based indicators apply at a virtual economic system level that contains all materials investigated. A great mix of materials already results in a certain leveling of various environmental characteristics since outliers lose their relevance [99]. This is valid far less at a micro level of product or process systems, e.g., in the case of paired material substitutions in processes or products, since the heterogeneity of environmental profiles of more similar materials is even more pronounced. An excellent example is the similarity in masses between copper and aluminum: copper meets multiple raw material requirements (in terms of CRD) and acidification potential, while aluminum has high energy demand and global warming potential (GWP)—which can be retraced to both mineralogical and technological factors. When substituting copper with aluminum, a decrease in raw material input and acidification potential comes at the expense of energy demand and greenhouse gas emissions.

In addition, mass-based input indicators do not reflect the environmental burdens or benefit of abatement technologies used within the economy. For example, very specifically mass-intensive materials in terms of raw material intensity may be accompanied by chemico-physical functionalities which can lead to significant resource efficiency gains and additional benefits in their material life cycle compared to technologies lacking these materials. Examples to be mentioned include platinum, palladium, and rhodium catalysts, which can significantly reduce the emission of nitric oxides and volatile hydrocarbons in combustion engines, although they themselves are among the most resource-intensive materials. A selective catalytic reduction system in diesel cars also represents an additional expense during the entire life cycle, but this leads to significant reductions in nitrogen oxides in exhaust gases. Scopes 3 and 4 should be selected according to Figure 6 as the most elaborate approaches in order to take account of the effects in the use phase. However, apart from the availability of data for such analyses, there is a great allocation problem as to which materials of a good we can attribute to the environmental impacts occurring in its further use. A UNEP study for the EU 27 and Turkey took Scope 3 according to Figure 6 using the actual annual material flows in terms of domestic material consumption [102]. In this case, the consumption phase was taken into account, which generally includes a consuming use of fossil fuels and leads to more intensive environmental pollution profiles. Very different characteristic environmental pollution profiles are obtained for various material categories such as fossil fuels or agricultural products. The respective material flows over different resource consumptions and impact categories do not follow a uniform pattern, which explains the barely existing correlations. This is one explanation that no linearity between material input into the economy and emissions out of the economy could be observed in these studies.

Only if the raw material use were lowered in equal measure for all raw materials, in other words, if a constant distribution of raw material use were assumed, would the overall environmental impact also be consistently reduced. However, this becomes generally invalid as soon as production and consumption patterns lead to deviating raw material input vectors (for example, by a material substitution from bricks to wood materials or simply technology changes regardless of the actual material input of a good). The actual variability of the macroeconomic material use and the cyclical dynamics in individual sectors over time, which can be traced back using environmental-economic accounting data [53,103], render this simplistic assumption seem unjustifiable.

Mass-based indicators are not good proxy indicators for relevant environmental impacts according to the studies presented. This makes establishing dematerialization problematic. If mass-based indicators alone are not very representative, the question arises whether they are important in a dashboard of indicators. A set of a small number of indicators can be determined to best describe environmental impacts through a system. This procedure can help reduce effort compared to a comprehensive assessment of all relevant impact categories by eliminating redundancy and thus increase relevance for decisions [104].

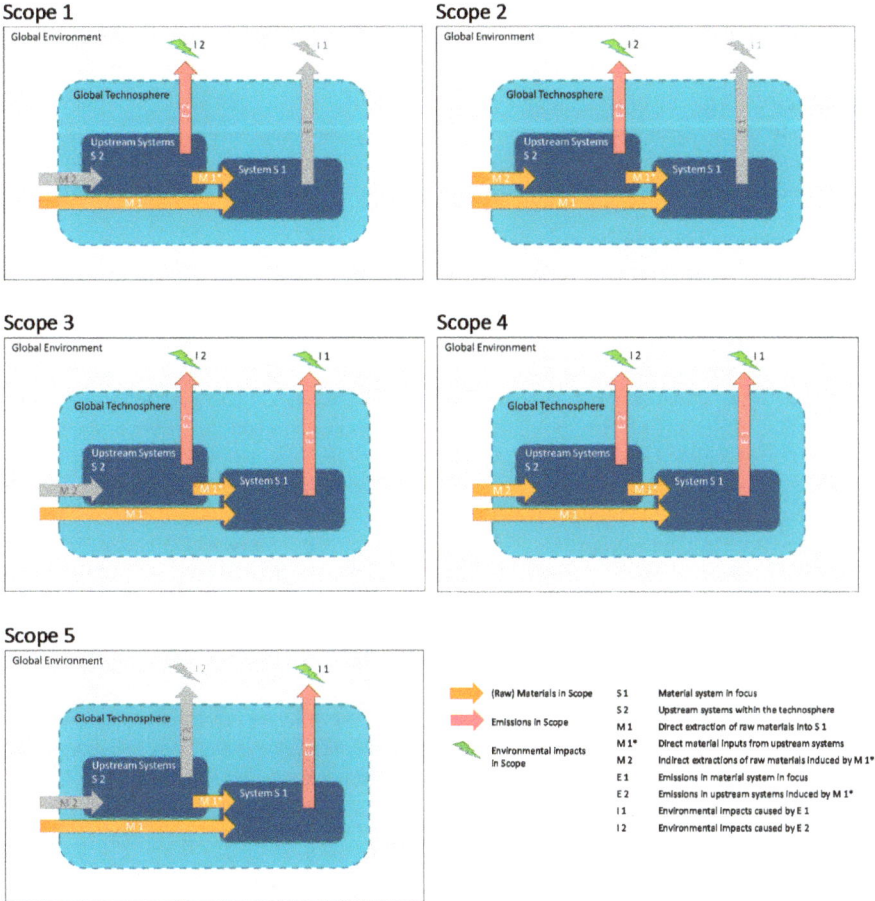

Figure 6. Two-tier raw material system to schematically show the different levels of impacts caused by a material input into System S 1 and its use. Scope 1: Direct material input (M 1 as direct extraction and M 1* as processed material from any upstream systems S 2) with environmental impacts I 2 in the upstream systems S 2; Scope 2: The sum of direct material input and the related indirect raw material inputs M 2 in the upstream systems S 2 with the environmental impacts I 2 caused there; Scope 3: Direct material input with the sum of the environmental impacts I 2 in the upstream systems and the environmental impacts I 1 caused in System S 1; Scope 4: The sum of direct material input and indirect raw material inputs M 2 in the upstream System S 2 with the sum of environmental impacts I 2 in the upstream systems S 2 and environmental impacts I 1 caused in System S 1; Scope 5: Direct material input with environmental impacts I 1 in System S 1.

An equally complex correlation analysis, a so-called principal component analysis of approximately 1000 raw materials, semi-finished, and finished products, plus 135 impact, midpoint, and end point indicators showed that the number of indicators can be reduced substantially while almost never losing significance [104]. A 92.3% variance between all 135 indicators could be explained by combining only six indicators—GWP, land use, stratospheric ozone depletion, marine ecotoxicity, terrestrial ecotoxicity, acidification, and eutrophication—in a highly representative set. In the case of environmental pressure indicators, 82% of the variance of all indicators can be explained by a combination of four variables: fossil CED, land use, water consumption, and raw material input. Thus, such a set of physical input indicators is not as reliable for use as a set as is a verifiable midpoint (impact) indicator set, but can nonetheless be a notable alternative. The fossil CED already represents 73%, which confirms the above-mentioned study results [97] as of equal importance to the best possible proxy indicator. As a comparison, the raw material input represents only 54% of the characteristics of all 135 indicators, which underlines its rather poor suitability and the availability of other, more appropriate alternatives. Using one of the proposed indicator sets will be the most robust option at a macroeconomic level. RMI or RMC as the only material flow indicator is of little significance.

3.2.2. Representation Function for Scarcities

The scarcity of raw materials is a keynote for resource conservation. Combined with the perception of large price fluctuations for raw materials, concerns are being raised that existing production patterns cannot be sustained on a permanent basis due to a critical scarcity of raw material-bound production factors that will occur on the input side. In the case of abiotic raw materials, the terminological distinction between actual raw material consumption ("depletion") and raw material availability is very important. From a geological point of view, the consumption concept refers to the depletion of a deposit or, at a global scale, to raw material resources combined in the lithosphere. However, availability is defined with regard to technical and economic conditions and is closely linked to market price trends and possible opportunity cost [105]. (Another availability term was introduced by Georgescu Roegen (The entropy law and the economic process. Harvard University Press: 1971). He applied the thermodynamic concepts of available and unavailable energy (entropy) to materials and argues that any kind of material is irrevocably degraded into unavailable states, e.g., by dissipation. For this reason, he concludes that the materials that are vital for current technologies in available states will become unavailable eventually. However, the reasoning considers the material transformation and losses within the economy and not just the use of the primary raw materials). Availability and consumption—expressed in terms of ranges—also refer to very different timelines, from a few years or decades up to many centuries in the case of the consumption of finite geogenic resources. Against this background, raw material scarcity discussions can be performed at three levels:

- physical depletion;
- structural supply risks;
- raw material criticality.

Physical Depletion

Measuring physical depletion is closely linked to the paradigm of geological fixed stock. The lithosphere provides a finite geopotential of raw materials. Neoclassical approaches to natural resource economics and the growth-critical positions taken by representatives of ecological economics—e.g., the influential "Limits to growth"—emphasize that raw material consumption exhausts the fixed stock that is assumed to be known [106,107].

A first rough orientation can be provided by the proportion of elements in the lithosphere (and the atmosphere for gaseous elements). Detached from short-term technological trends and demand situations, this enables a categorization of the elements that are available to the extent that they cannot be exhausted in human time periods. Accordingly, the elements that occur in typical rock formations—silicon, aluminum, iron, calcium, sodium, potassium, and magnesium—all of which are present in the single-digit percentage range and together with oxygen and hydrogen form 98% of the lithosphere [108] are also referred to as *Elements of Hope* [109]. In addition to this categorization, elements are also grouped as scarce and critical elements. When considering these *Elements of Hope* together with carbon, which constitutes approximately 80% of fossil fuels and 50% of biomass, it can be estimated by the example of Germany and the EU-28 that more than 99% of RMI is characterized by these 10 elements [52,103]. Approximately 30% of this is carbon. Accordingly, macroeconomic material flow indicators also significantly represent the elements that are physically most common on Earth.

The geological content of elements does not in itself indicate any economic or technical extractability. If the top-down approach of the fixed stock is to be adhered to, the geochemical postulate of the mineralogical barrier stipulates an average threshold for all elements (particularly relevant for scarce and critical elements), which represents the stoichiometric limit concentration in ores. Below this threshold, most of these elements are dispersed thinly in rock, such that extraction could only be achieved through wet-chemical separation of the rock rather than by separating accumulated fractions [110,111]. Even if the postulate of the mineralogical barrier is not scientifically proven and is indirectly questioned by reference to the exploration cycles and technological advances in the mining industry, it still prompts a fruitful discussion about the increase in extraction costs [112–114]. This looks into the topic of resource intensity of extraction, particularly with regard to energy demand and the resulting emissions. However, this "environmental scarcity of raw materials" is revealed by other tangible AoPs, not the physical scarcity of extractable raw material amounts per se. Material flow indicators do not have any direct or indirect links in this context.

Only a few exceptions present any empirical evidence for absolute, static, and physical scarcities—in resource economics, the complete exhaustion of a resource or a raw material. Peak oil is one of these exceptions. The peak oil phenomenon predicted in the 1950s by Hubbert [115] for US oil production and in later projections for global production as a whole—described in light of an analysis of new discoveries of conventional oil deposits—is widely regarded as being substantiated [116]. However, a transfer of Hubbert's peak analysis concept with the assumption of a fixed-stock of mineral raw materials, as occasionally propagated [112,117,118], is regarded by many authors as critical [119,120]. This is because the concept does not take into account the actual causes of supply and demand development for mineral, non-fossil raw materials, nor does it consider the variability of the reserves and resources parameters [121]. Criticism is directed particularly at the static lifetimes that are misinterpreted as lifetime up to the exhaustion of a material [122]. Retrospectively, it appears that the static lifetimes have remained almost constant or even increased for both fossil energy resources and ores in spite of a conspicuously increased consumption development [123] in the second half of the 20th century [124]. The fact that the reserve range of rare earth metals is around 1048 years, that for copper is around 40 years, and that for lead is only 20 years is not indicative of an alleged earlier exhaustion of global lead deposits [125]. Regardless of this limited significance of static lifetimes for physical scarcities, they cannot be derived from aggregated mass-based flow indicators.

Structural Supply Risks

In addition to physical depletion, technical-structural aspects that can limit global availability of raw materials are also discussed. Some of them can be objectified and are assessed independently from the players and their use. This method takes into account spatial disparities of distribution, and the resulting geographical concentrations of the deposits, and of raw material production [126,127]. Due to geological and geochemical conditions, the most significant known deposits of some metals, such as tin (soap deposits in Indonesia), niobium (Brazil), platinum (South Africa and Zimbabwe), palladium (Ni-Cu deposits around Norilsk/Russia), and cobalt (Central African copper belt, especially in DRC), accumulate in a few districts worldwide. Others—such as rare earths—do not occur in such a high geological concentration, but in a sizeable concentration in the current mining production (China). Country and enterprise concentrations in an oligopolistic market environment have in the past repeatedly led to sensitive supply disruptions with high price fluctuations [128].

A further basal structural supply risk affects metals that are present but associated with other materials in their ores. Many metals can only be extracted as by-products or co-products. Exploration, production volumes, and related ranges are thus also bound to the carrier metal [122,129]. By-products are present in comparatively lower concentrations and their availability is strongly or exclusively coupled to the production capacity of one or more carrier elements, for example, gallium in bauxite ores for aluminum production as well as indium and germanium from zinc and lead production. On the other hand, co-products have no clear carrier element. The metals concerned such as platinum group metals (PGMs) and rare earths exist in defined mineralogical ratios, which require joint processing and refining in order to purify the individual elements. Due to their large ionic radii, rare earths are not finely dispersed in typical rock formations, but form their own minerals in which they are jointly integrated in characteristic ratios [108]. Regardless of the element-specific demand, actual production is determined by the characteristics of the deposits. As a result, there are production-driving elements—formerly europium and cerium, currently neodymium and dysprosium—that determine the production volume and others whose production volume exceeds demand, which means that a low market price is highly likely due to the overproduction of such elements, for example, cerium and lanthanum [130].

The outlined structural risks are to be regarded as element-specific and require a geoscientific statistical model. Conclusions cannot be drawn about the rise of structural supply risks using material flow indicators, even in disaggregation, without this complementary model knowledge.

Raw Material Criticality

The assessment of raw material criticality goes beyond the two previous scarcity scales in two respects. On one hand, supply risks are more complex. In addition to the geological, technical, and structural criteria, geopolitical, socio-economic, and ecological risks are also integrated [93,131]. On the other hand, the importance of raw materials (and their uses) for a specific system and the impacts in the event of a manifested supply risk are taken into account.

Against this background, criticality analysis aims to identify the raw materials of a system that uses them (for example, national economy, industry, and enterprises) to fulfill essential functions for this system but whose supply is at risk. This enables a vulnerability assessment of that system in the face of supply disturbances of specific raw materials [93]. The criticality assessment does not correspond to the characterization of total, but relative scarcities, which occur when raw material demand of the reference system cannot be satisfied by the raw material supply in terms of time, space, and organization. Linked to this, the impact horizon can also have global, continental, and national impacts, or may only be (supra-) regionally limited.

Such scarcities can occur due to a dynamic supply or demand development as well as a combination of both. The more a raw material suffers from a supply risk and the more inelastic its demand will be, the more critical will it become to the reference system [93]. Relative scarcities must therefore also be assessed depending on the parties involved; for example, diversification of raw material suppliers or internal recycling can reduce the relative scarcity risk of a single market participant.

The first studies that correspond to today's logic of criticality analysis were motivated by a raw material security policy and date back to the 1970s [132–134] and the basics even to the 1950s. Numerous new criticality metrics have been developed since 2008 [135]. Different raw materials are now classified as critical depending on the criteria used within these metrics, the reference systems, and the up-to-dateness of the analyses. Discrete lists of critical and uncritical raw materials—e.g., 20 out of 54 raw materials of an ad hoc working group of the EU COM [136]—can create a strong impact signal in communication. However, it also counteracts the relative nature of the concept [121].

To allow for a universal and flexible application of the concept on different levels from macro- to micro-scales, i.e., countries, economies, regions, industries, branches, sectors, technologies, companies, product lines, products, and building blocks, an expert group at the Association of German Engineers (VDI) has elaborated a harmonized criticality methodology as part of an industrial guideline on resource efficiency (VDI 4800-2) [93]. The two-dimensional approach clearly distinguishes between exogenous geological, technical, structural, geo-political, regulatory, and economic criteria to characterize the overall supply risks and subjective endogenous vulnerability criteria for the object in focus. Current work focuses pragmatically on the future. Thus, the expected demand impulses due to dynamically emerging technologies are systematically estimated [123,137,138] or the impacts of climate change on the future supply of raw materials are analyzed in risk assessments [139,140]. The results can be integrated in prospective criticality analyses. Other research initiatives do not end with the assessment of criticality but develop approaches to mitigate it by efficiency, recycling, or planned substitution measures [141,142]. These strategy initiatives are aimed at both the supply and the demand side and are not generic, but they are explicitly aligned to the respective technological functional materials. Moreover, the assessment of ecological and social impacts of raw material supply is further developed in criticality analyses [143–147]. This has not been systematically taken into account in existing criticality assessment methods and forms a link to sustainability assessment [148]. The example of coltan mining in the Democratic Republic of Congo showed in the past that strategic importance of raw materials, in this case tantalum, can also aggravate and catalyze armed conflicts in producer countries [149,150]. In addition, overuse of ecosystems and degradation of further natural resources by the extraction of raw materials can lead to violent conflicts [124,147,151,152].

In the trend of current studies, raw materials for environmental and emerging technologies are classified as particularly critical. These include, for example, essential raw materials for thin-film photovoltaic technologies such as indium, tellurium, gallium, and germanium. In addition to these specialty metals, precious metals such as platinum, rhodium, and gold, or refractory metals required in significantly larger amounts such as tungsten, molybdenum, tantalum, niobium, and chromium, are also classified as critical or, from a national perspective, as "economically strategic" [141].

At a global scale, total use of raw materials rose from about 40 billion tonnes to 75 billion tonnes between 1993 and 2013 [153]. With its growth factor of approximately 1.9, it has grown slightly stronger than the global economic performance (an increase factor of 1.75 with an average annual growth of 2.8%). Thus, there is a minor relative decoupling of the raw material input from the GDP at the aggregated raw material level over the long term. Although significant demand increases are still expected for many materials considered as critical, their production—including germanium, indium, tantalum, and cobalt—has multiplied between 2 to 5 times in the same 20-year period [123]. Consequently, their use is clearly decoupled from the use of the total primary raw materials, which indicates that broad-brush approaches for dematerialization do not solve criticality issues at all.

Criticality analysis has established itself as a multi-layered and complex, action-oriented socio-economic raw materials assessment method to deal with scarcities. Thus, it is a highly relevant method to address the safeguard subject raw material availability. In principle, intra- and intergenerational availability and distribution fairness can be derived as socio-economic AoPs within the criticality concept (see Section 3.1). It could be argued, taking an egalitarian value scale, that critical functional materials should be available to the same extent as a resource for the fulfilment of a benefit both intra- and intergenerationally. However, conclusions cannot be derived in general—only discreetly linked to the elements and materials investigated. This action field cannot be addressed using a static development of material flow indicators simply because most critical raw materials link to technological and specialty metals plus industrial minerals that have relatively small mass flows and can barely be dealt with in aggregated mass flow calculations at this depth. A general dematerialization does not lead to the reduction of criticality of raw materials. Instead, more focused complementary material flow analyses are required that also take into account the inventory dynamics of product groups and functional materials bound in them [154].

3.3. Exegese of the Precautionary Principle

Mass-based indicators failed to provide a strict representation for either environmental impacts or scarcity. One reason may be that qualitative and quantitative relationships and cause–effect relationships are not sufficiently known or are not mapped by the methods discussed in this article. Some authors, who propagate dematerialization as the central and dominant maxim of action for resource conservation [59], borrow the precautionary principle as a context of justification from environmental policy. It serves both the derivation for the basic necessity of a reduction of primary material extraction and a determination of different quantitative targets of dematerialization [30]. Since the precautionary principle is at the core of justification for dematerialization strategies (see Section 2.3), we take a closer look at this central principle of environmental policy in this section.

The European Union's environmental policy is based on the precautionary principle (Article 191 of the Treaty on the Functioning of the European Union). Internationally, the precautionary principle was enshrined in the 1992 Rio Declaration on Environment and Development [155] as a principle for the protection of the environment and human beings. The precautionary principle guides environmental policy towards acting early and with foresight to avoid future environmental pressures or preserve natural resources for future generations, even if the knowledge of type, extent, likelihood, and cause–effect relationships is incomplete or uncertain.

Despite its importance in the political decision-making process, there is always confusion as to what the principle actually means and in what cases it can be applied as a principle of action. In the case of dematerialization as a strategy for resource conservation three main conditions and one assumption can be initially identified (see Section 2.3): The knowledge about exceeding the planetary boundaries is incomplete and uncertain [1]. There is an extensive lack of knowledge about cause–effect relationships in the environment. Estimations suggest that man-made material flows exceed the extent of natural material flows. With reference to the laws of conservation of mass and energy, it can be hypothesized that the material input into the socio-economic-system could be a proxy for the total environmental impact potential of this system (see Figure 6) [156]. However, do these arguments suffice to legitimize a dematerialization agenda and formulate dematerialization targets making reference to the precautionary principle?

The European Commission has issued a communication on the applicability of the precautionary principle (COM (2000) 1) [157]. According to the European Commission, the precautionary principle can legitimize or even impose political acts in such cases where scientific uncertainty is high, but there is concern due to a scientific risk assessment that dangerous consequences for humans and the environment are inconsistent with the EU's high protection standards. However, the precautionary principle can under no circumstances be used to justify the adoption of arbitrary decisions. In order to apply the precautionary principle, the following prerequisites must be met, unless there is a concrete perception of risk, according to the EU Commission: Before the precautionary principle is invoked, potentially negative effects of a phenomenon, must be identified based on the relevant scientific data and should be scientifically examined. When considering whether measures are necessary to protect the environment, every effort should be made to scientifically evaluate the available information. Reliable scientific data and logical reasoning are required to come to a conclusion, which expresses the possibility of occurrence and the severity of a hazard's impact. If the available data is sufficient, scientific uncertainties must be determined and assessed at each stage of the procedure. Do these prerequisites apply in the case of dematerialization?

The dematerialization logic of the factor X concept postulates that the material input measured in mass units is causally related not only to the environmental impacts of the upstream chains of material production but also to the total environmental impacts of the system in which the material is used (see Scopes 3 and 4 in Figure 6). The postulated connection between mass and environmental impact potential is based solely on plausibility arguments [156]. There is still no scientific evidence (see Section 3.2.1 and [30]). Regression analyses failed to show either a link between direct and indirect material demand and environmental impacts caused in upstream systems, or for the direct material consumption of a system and the environmental impacts generated by this system. An explanation for the first case is the heterogeneity of the environmental profiles of the materials used, which is not taken into account by the mere addition of the masses. For the second case, an explanation approach is that the material input into a system does not map the functions of the materials within the system and the way the materials contribute to environmental impacts or benefits (see Section 3.2.1).

However, the empirical evidence that has not been provided does not yet preclude the application of the precautionary principle. On the contrary, it is precisely in this case that legitimation of environmental protection measures can be considered if the presumed cause–effect relationships are plausible and can be described based on reliable scientific data and logical reasoning. However, such a description that allows a conclusion on the occurrence and severity of a hazard's impact is not available. In contrast to the application of the precautionary principle in other environmental policy areas, e.g., in the field of potentially dangerous substances, the cause–effect relationships cannot be systematically described biophysiochemically in the case of dematerialization logic. Although it is plausible that overall mass input must eventually equal overall mass output, the entire material input into a system cannot be brought into a logical cause–effect relationship with the environmental impacts exerted by this system. For small systems such as products, the large allocation problem mentioned in Section 3.2.1 should be solved, that is, the impacts of the product along its life cycle ought to be attributed to the materials entering the product system. Although, for example, the reference to greenhouse gas emissions can be produced in the case of fuels, nitric oxide emissions cannot be attributed solely to the fuel, but are a function of combustion technology and exhaust gas purification technology, which require other raw materials. This reference can hardly be established by logical reasoning for constructive or functional materials used, for example, in electronics. In larger, dynamic systems such as economies, a modeling of the cause–effect relationships is an almost impossible undertaking due to the multiplicity of factor combinations in the use and further processing of materials into products and the environmental benefits and environmental impacts of the same products along their entire life cycle, user behavior, and product design decisions. An assessment of the scientific uncertainty is only possible if these fundamental allocation and modeling problems are solved.

Therefore, most of the EU Commission requirements on the applicability of the precautionary principle are not met. According to these criteria, resource conservation or efficiency policies focusing on mass-based indicators and having the objective of reducing environmental pressure cannot be legitimized by the precautionary principle. Setting global dematerialization as well as material efficiency targets is, therefore, arbitrary within the meaning of the communication [157], and the precautionary principle cannot be relied upon. Bringezu also argues that dematerialization targets should be set "rather arbitrarily" due to the lack of scientific evidence "void of better alternatives" and should be based on Schmidt-Bleek's proposals [30,59]. However, there is an alternative to general dematerialization by an arbitrary factor as a strategy for the urgently needed response to the threat of exceeding planetary carrying capacity. The complexity of socio-economic systems and the planetary ecosystem that oppose the scientific check of the "Schmidt-Bleek conjecture" can be reduced by focusing on the impacts of socioeconomic systems and breaking the planetary ecosystem down into a few large compartments. This has been shown by the work of Rockström et al. on the planetary boundaries [1,3,158]. Although his derivation of the planetary boundaries is also scientifically controversial, it is based on scientific and at least partly empirically provable cause–impact relationships, which, according to EU-COM (2000) 1, qualifies it for the legitimation by the precautionary principle and thus as a measure of action for environmental policy. Even though, the disaggregation of planetary boundaries seems to be more promising for justifying the precautionary principle, whether all relevant environmental impacts and social conflicts in the nexus of land use and raw material extraction are reflected has to be checked.

4. Findings

This paper critically assessed the relevance and representativeness of mass-based indicators' multiple layers as a basis for resource policy. The current application of material flow indicators and targets was discussed, and concepts for raw materials and safeguard subjects were juxtaposed from a micro-level such as LCA. Thereof, conclusions are drawn about the robustness of dematerialization as a strategy for resource conservation.

- Mass-based indicators that consider primary raw materials as objects of balance and consequently trace back all material flows entering a system by raw material equivalents (RMEs) have the greatest significance since they unambiguously map a specified natural resource, the primary raw materials. They explicitly address the AoP natural resources in terms of raw material consumption for biotic and abiotic raw materials. An intrinsic value is attributed to raw materials, expressed by its mass, and any form of raw material extraction and removal ex aequo impairs the AoP natural resources. This does not apply to narrow indicators that only consider direct material input or even more complex indicators that not only include indirect raw material flows in upstream systems but also hidden material flows (unused extractions). In the case of the latter ones, there are asymmetries in balancing and excessive data uncertainties.

- However, raw-material, mass-based indicators to which qualitative relevance has been attributed also fail to analytically map other resources. It is not tenable to draw a conclusion about high "resource relevance" from high physical material or raw material inputs. One can at most speak of "raw material relevance".

- All aggregated mass-based material flow indicators including those in RME have at best a weak representation function for negative environmental impacts. According to correlation analyses, input indicators for raw materials and materials do not qualify as provable significant parameters concerning either environmental impacts of raw material production for a socio-economic system or the total environmental impact of this socio-economic system. If the representation function is already rudimentary at a macroeconomic level, it is even weaker at the micro level. Basically, the smaller and more fine-grained the system is, the less individual indicators will be predestined as representatives. System optimization based on material input can lead to significant increases in other resource consumptions during production or further uses over the life cycle. An obvious consensus prevails in the literature that non-renewable primary energy demand as a physical input quantity in units of energy is the best possible solitary representative of a system's further environmental impacts.

- In light of methods and findings from the assessment of raw material availability and consumption, no representation function for avoiding raw material scarcities is available either for absolute, structural, or relative socio-economic scarcities, i.e., criticalities. No general and plausible relationships can be constructed using aggregated material flow indicators for raw material scarcities. On the other hand, this requires a strong material-exact disaggregation. Nonetheless, the criticality concept complements the resource efficiency concept by a relevant aspect: the significance of the raw materials used for a system. The specific criticalities yield important reference points for the prioritization and derivation of effective measures and tools to increase the efficiency of their use, recycling, or substitution.

- Justifying dematerialization strategies and target setting with reference to the precautionary principle is controversial since it ignores the lack of empirical representativeness of mass-based material flow indicators for known environmental impacts. In addition, the cause–effect relationships between mass-based material input into a system and the known and unknown impacts of a system cannot be reasonably produced using logical reasoning due to major allocation problems and the complexity of large systems such as economies, which also makes it impossible to assess the scientific uncertainties.

5. Conclusions

While complementary mass-based indicators, which describe the socio-economic metabolism, provide valuable information for resource policy, there is little evidence to consider solitary physical, aggregated mass-based indicators as control and target variables for achieving a strategy for resource conservation. For many years, dematerialization representatives have emphasized the merits of the ability to apply indicators of the same systematics at all levels of the socio-economic system and to base design, act, and optimization on them. Although the laws of conservation of mass and energy indicate that more input eventually must result in more output, the postulated link between the overall material inputs into a system and its outputs—not in terms of waste generation, but in terms of resource consumption and environmental impacts—is not demonstrable generically or backed by any thermodynamic foundation. This assumption is scientifically controversial, since it is based on barely comprehensible logical reasoning and lacks a sound empirical foundation. Dematerialization for its own sake is barely convincing and should not be considered as an end in itself It is only reliable as a strategy for resource efficiency enhancement or resource conservation under almost constant, unrealistic conditions in terms of the qualitative input factors: Only if macroeconomic raw material use was lowered ceteris paribus for all raw materials could the environmental impact be reliably reduced to the same extent. However, this becomes invalid as soon as production and consumption patterns change and lead to deviating raw material use vectors. The actual variability of the macroeconomic material use and the cyclical dynamics in individual sectors over time make this simplistic assumption seem unjustifiable.

At the level of individual processes or products, such an approach using indicators such as MIPS is highly questionable. Moreover, for systems as large as national economies, no stringent design basis is obtained from the size of absolute raw material inputs or material uses. In the opinion of the authors, the main reason is not a missing life-cycle perspective such as neglecting upstream flows beyond a national state boundary (see Sections 2.2 and 3.2.1). It is rather the variety of options for factor combinations, processing technologies, and utilization of raw materials combined with a vast portfolio of possible abatement technologies and waste management practices that determine the environmental and resource implications of a material system of any size.

On the one hand, mass-based indicators are assigned a strong communication function in an information hierarchy and might be the best available indicators for reporting purposes. On the other hand, no conclusions can be drawn about raw material-related AoPs beyond the intrinsic value of the raw materials. To give an example, the extraction of bulky building materials such as sand, gravel, and clay may be of minor significance in Germany and countries alike that, compared to India or Bangladesh, where severe environmental and social conflicts are evoked, have rather complex licensing procedures [159]. This is problematic because it does not provide a convincing orientation framework for dematerialization as a resource conservation strategy. Since there is no reason to believe that dematerialization generally contributes to the conservation of universally accepted guiding principles and AoPs, or that it prevents adverse effects, it is vehemently rejected by many societal political and economic groups, who fervently turn against any form of absolute cap in production and consumption. The vague invocation of the precautionary principle is argumentatively too weak to be convincing, yet the same stakeholders advocate an increase in resource efficiency as a priority strategy [160].

Even if dematerialization is hardly consensual, its narrative—the over-use of the natural resources of the Earth and achieving or already transcending planetary boundaries—receives increasing approval. However, rather than indulging too much in the absolute material and raw material inputs and throughputs of society that cannot represent this degradation and further relevant aspects of resource use, more adequate and diverse indicators and methods can be selected. The level of detail must be increased and the functional resource categories considered must be broadened (see Section 2.1) to be able to derive reliable conclusions on resource efficiency. As mentioned in Section 3.2.1, there is evidence that the multitude of natural resources might be represented by a reasonable number of equal resource indicators in a set without the need for a hardly consensual normative step of weighting.

Despite the limitations of aggregated material indicators and data, mass-based information of a system can be the starting point for essential information using disaggregation, standardization, and the application of characterization models. The material-based approaches themselves thus provide the basis for further, widespread methods, which can be placed in a systematic context, e.g., by applying hybrid MFA-LCA approaches [54]. In concluding the assessment made in this paper, the authors plead for a systematic and concise use of the different methods and notions as proposed and explained in Figure 7 in resource policy instead of a simplistic dematerialization focus.

The figure shows the logical and semantic linkage of the utility value, the availability as well as the natural resources with assessment approaches to material efficiency, resource intensity, supply risks, criticality, resource efficiency, resource consumption, and sustainability in a material-economical system. The focus is on material information. None of the assessment approaches can be performed using solely the aggregated mass as it is recorded in mass-based indicators. However, these targeted, characterized, and impact-oriented approaches are imperative for consistent resource conservation and need further development. In particular, the measurement of resource consumption hits the core of controversial debates in life cycle impact assessment (LCIA) about the selection of adequate impact categories, indicators, impact pathways, characterization models, and optional weighting and aggregation steps, for which consensus building is an important and urgent matter.

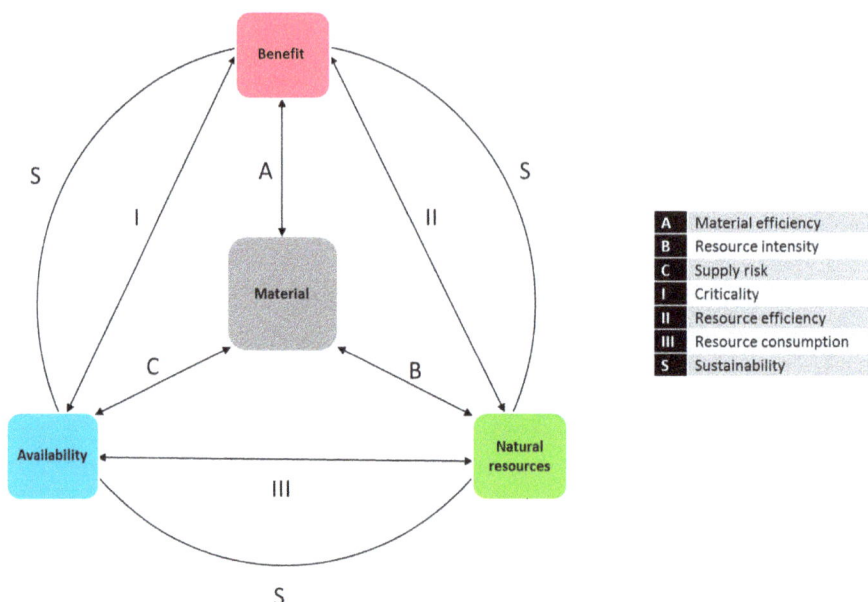

A	Material efficiency
B	Resource intensity
C	Supply risk
I	Criticality
II	Resource efficiency
III	Resource consumption
S	Sustainability

Figure 7. Relation of material flow information and determination methods for resource conservation. If a material inventory (**Material**)—such as a product's bill of materials (BoM)—was determined for a system using material flow analysis, substance flow analysis or inventory analysis methods, this can be related to a benefit in terms of a technical utility value or a monetary success rate of the product in order to determine its material efficiency (**A**). Coefficients for resource use (**Natural Resources**) enable the determination of the resource intensities (**B**) (for primary raw materials, fresh water, land, greenhouse gases, etc.) up to the provision of the product (as amount/number) from the material inventory. Using socio-economic data on the materials used (**Availability**), it is possible to determine the material's supply risks (**C**). Considering the significance of the product for a reference system, for example, a manufacturing company, and the supply risks together enables a criticality analysis (**I**). Similarly, the resource efficiency can be determined from the benefit of the product for the reference system and the resource intensities (**II**). The actual resource consumptions (**III**) can be balanced based on resource use in connection with the supply situation (regional water availability, land quality, etc.) (**Availability**). Thus, Availability refers to socio-economic and natural occurrences. If all approaches are integrated and dynamized, questions of intra- and intergenerational distribution and resource fairness can be investigated, which will lead to a sustainability assessment (**S**).

As Mancini et al. suggest, the choice of accounting methodology and the setting of targets should be guided by policy objective, be it reducing environmental pressure, enhancing resource availability, or maximizing benefit from natural resource use, not the other way around, which seems to be the case for dematerialization approaches according to our assessment and findings [54]. Although numerous environmental policies concentrate on pressures and impacts already, adding an impact perspective to the so far widely material-focused resource policies would create an additional benefit. Most environmental policies lack a life-cycle perspective and focus on emissions and wastes within national state boundaries. They neglect upstream pressures and impacts associated with imported goods and, by doing so, enable burden shifting effects which need to be considered when assessing the impacts of consumption and production. Uncovering burden shifting is, however, a strength of material flow accounting. Combining the two approaches would be another step towards a more complete and robust accounting of natural resources and its use [161].

The rigorous conceptual development of resource conservation as a policy and science field not only must solve the conceptual node in relation to environmental protection (see Section 2.1) and avoid equating material and natural resources, but also has to answer the question of the relevant target parameter of resource conservation: If the goal of the sustainable use of natural resources proclaimed in the SDG had been achieved, there would be no need to further reduce resource use. On the other hand, resource consumption, in the sense of a consuming use of natural capital in the context of local, regional, and planetary ecological sustainability limits, is the relevant target parameter in this respect that must be decoupled from human well-being in the sense of absolute decoupling. However, this has not yet been operationalized using indicators in any of the policy strategies mentioned at the outset in Section 1.

Acknowledgments: The authors would like to acknowledge everyone who contributed to the controversial discussion on the topic at the German Environment Agency and the technical support for submission provided. The suggestions and insights from the anonymous reviewers are greatly appreciated. The opinions expressed in this article are the authors' own and do not necessarily reflect the views of the German Environment Agency.

Author Contributions: The manuscript was conceived, elaborated, written, and edited for submission by F.M. and J.K. Thoughtful discussion and critical review was provided by H.K., M.A. and B.R.

Conflicts of Interest: The authors declare no conflict of interest.

Abbreviations

The following abbreviations are used in this manuscript:

AoP	area of protection
CED	cumulative energy demand
CRD	cumulative raw material demand
DMC	domestic material consumption
DMC_{abiot}	abiotic domestic material consumption
DMI	direct material input
DMI_{abiot}	abiotic direct material input
LCIA	life cycle impact assessment
MIPS	material intensity per service unit
RME	raw material equivalents
RMI	primary raw material input
RMI_{abiot}	abiotic primary raw material input

References

1. Rockström, J.; Steffen, W.; Noone, K.; Persson, A.; Chapin, F.S.; Lambin, E.F.; Lenton, T.M.; Scheffer, M.; Folke, C.; Schellnhuber, H.J.; et al. A safe operating space for humanity. *Nature* **2009**, *461*, 472–475. [CrossRef] [PubMed]
2. Faulstich, M.; Foth, H.; Calliess, C.; Hohmeyer, O.; Holm-Müller, K.; Niekisch, M.; Schreurs, M. *Environmental Report 2012: Responsibility in a Finite world*; Erich Schmidt Verlag: Berlin, Germany, 2012; p. 716.

3. Steffen, W.; Richardson, K.; Rockström, J.; Cornell, S.E.; Fetzer, I.; Bennett, E.M.; Biggs, R.; Carpenter, S.R.; de Vries, W.; de Wit, C.A.; et al. Planetary boundaries: Guiding human development on a changing planet. *Science* **2015**, *347*, 1259855. [CrossRef] [PubMed]

4. European Commission. *Thematic Strategy on the Sustainable Use of Natural Resources*; (COM 670 (2005)); European Commission: Brussels, Belgium, 2005.

5. European Commission. *Roadmap to a Resource Efficient Europe*; European Commission: Brussels, Belgium, 2011.

6. European Commission. *A Resource-Efficient Europe: Flagship. Initiative under the Europe 2020 Strategy*; European Commission: Brussels, Belgium, 2011.

7. Kazmierczyk, P.; Geerken, T.; Bahn-Walkowiak, B.; Vanderreydt, I.; Veen, J.V.; Veneziani, M.; De Schoenmakere, M.; Arnold, M. *More from Less: Material Resource Efficiency in Europe*; 2015 Overview of Policies, Instruments and Targets in 32 Countries; Publications Office of the European Union: Luxembourg, 2016; Volume 10.

8. Federal Ministry for the Environment, Nature Conservation and Nuclear Safety (BMU). *German Resource Efficiency Programme (ProgRess)—Programme for the Sustainable Use and Conservation of Natural Resources*; Federal Ministry for the Environment, Nature Conservation and Nuclear Safety (BMU): Berlin, Germany, 2012; pp. 1–116.

9. Federal Ministry for the Environment, Nature Conservation, Building and Nuclear Safety (BMUB). *German Resource Efficiency Programme II (ProgRess II)*; Programme for the Sustainable Use and Conservation of Natural Resources. Progress Report 2012–2015 and Update 2016–2019; Federal Ministry for the Environment, Nature Conservation, Building and Nuclear Safety (BMUB): Berlin, Germany, 2016; pp. 1–158.

10. Government of Japan. *Fundamental Plan for Establishing a Sound Material-Cycle Society*; Ministry of the Environment: Tokyo, Japan, 2003; pp. 1–12.

11. National Institution for Transforming India Aayog. *Strategy Paper on Resource Efficiency*; National Institution for Transforming India: New Delhi, India, 2017; pp. 1–45.

12. G7/G20 Sherpa Office of the Federal Chancellery. *Final Report by the Federal Government on the G7 Presidency 2015*; Press and Information Office of the Federal Government: Berlin, Germany, 2015.

13. European Commission. G20. In Proceedings of the Leaders' Declaration—Shaping an Interconnected World, Hamburg, Germany, 7–8 July 2017.

14. United Nations. General Assembly. In *Transforming Our world: The 2030 Agenda for Sustainable Development*; Resolution Adopted by the General Assembly on 25 September 2015; A/RES/70/1; United Nations: Washington, DC, USA, 2015; pp. 1–35.

15. Rizos, V.; Tuokko, K.; Behrens, A. *The Circular Economy: A Review of Definitions, Processes and Impacts*; Research Report No 2017/8; Customs Excise and Preventive Service: Brussels, Belgium, 2017; pp. 1–44.

16. EU COM. *Closing the Loop—An Eu Action Plan for the Circular Economy*; European Commission: Brussels, Belgium, 2015.

17. Sitra. *Leading the Cycle: Finnish Road Map. to a Circular Economy 2016–2025*; Sitra Studies 121; Sitra: Helsinki, Finland, 2016; p. 56.

18. Government The Netherlands. *A Circular Economy in the Netherlands by 2050*; The Ministry of Infrastructure and the Environment and the Ministry of Economic Affairs, also on Behalf of the Ministry of Foreign Affairs and the Ministry of the Interior and Kingdom Relations: Hague, The Netherlands, 2016; p. 72.

19. Reichel, A.; Schoenmakere, M.D.; Gillabel, J. *Circular Economy in Europe: Developing the Knowledge Base*; Publications Office of the European Union: Luxembourg, 2016; Volume 2.

20. *Report of the Inter-Agency and Expert Group on Sustainable Development Goal Indicators*; United Nations Economic and Social Council: New York, NY, USA, 2016; pp. 1–62.

21. Federal Statistical Office (Destatis). *Nachhaltige Entwicklung in Deutschland—Indikatorenbericht 2016*; Federal Statistical Office (Destatis): Wiesbaden, Germany, 2017; pp. 1–150. (In German)

22. Wiedmann, T.O.; Schandl, H.; Lenzen, M.; Moran, D.; Suh, S.; West, J.; Kanemoto, K. The material footprint of nations. *Proc. Natl. Acad. Sci. USA* **2015**, *112*, 6271–6276. [CrossRef] [PubMed]

23. Demurtas, A.; Sousanoglou, A.; Morton, G.; Humphris-Bach, A.; Essig, C.; Harding, L.; Cole, A. *EU Resource Efficiency Scoreboard 2014*; European Commission: Brussels, Belgium, 2015; pp. 1–68.

24. The Organisation for Economic Co-operation and Development. Material flow and resource productivity indicators. In *Working Party on Environmental Information*; OECD: Paris, France, 2012; pp. 1–24.

25. Schmidt-Bleek, F.; Bierter, W. *Das MIPS-Konzept : Weniger Naturverbrauch—Mehr Lebensqualität durch Faktor 10*; Droemer Knaur: München, Germany, 1998. (In German)

26. Adriaanse, A.; Bringezu, S.; Hammond, A.; Moriguchi, Y.; Rodenburg, E.; Rogich, D.; Schütz, H. *Resource Flows: The Material Basis of Industrial Economies*; World Resources Institute: Washington, DC, USA, 1997.

27. Matthews, E.; Amann, C.; Bringezu, S.; Hüttler, W.; Ottke, C.; Rodenburg, E.; Rogich, D.; Schandl, H.; Van, E.; Weisz, H. *The Weight of Nations-Material Outflows from Industrial Economies*; World Resources Institute: Washington, DC, USA, 2000; pp. 1–135.

28. Kågeson, P. *Is Factor 10 a Useful Tool in Environmental Policy?* Swedish Environmental Protection Agency: Stockholm, Sweden, 1999; pp. 1–31.

29. Kleijn, R. Adding it all up. *J. Ind. Ecol.* **2001**, *4*, 7–8. [CrossRef]

30. Bringezu, S. Possible target corridor for sustainable use of global material resources. *Resources* **2015**, *4*, 25–54. [CrossRef]

31. Reijnders, L. The factor x debate: Setting targets for eco-efficiency. *J. Ind. Ecol.* **1998**, *2*, 13–22. [CrossRef]

32. Wiedmann, T.; Minx, J.; Barrett, J.; Vanner, R.; Ekins, P. *Sustainable Consumption Und Production—Development of an Evidence Base*; Project Ref.: SCP001 Resource Flows; Stockholm Environment Institute: Stockholm, Sweden, 2006; pp. 1–133.

33. European Economic Area. Gemet—Environmental Thesaurus: Term Natural Resource. Available online: https://www.eea.europa.eu/help/glossary/gemet-environmental-thesaurus/natural-resource (accessed on 2 October 2017).

34. Enquete-Kommission Schutz des Menschen und der Umwelt. *Die Industriegesellschaft Gestalten—Perspektiven für Einen Nachhaltigen Umgang mit Stoff—Und Materialströmen*; Deutscher Bundestag: Bonn, Germany, 1994. (In German)

35. United Nations. *Convention on Biological Diversity*; United Nations: New York, NY, USA, 1992.

36. Nantke, H.-J.; Gottlob, D.; Summerer, S.; Burger, A.; Dieter, H.H.; Friedrich, A.; Fritz, K.; Greiner, P.; Hahn, J.; Henseling, K.O. *Nachhaltige Entwicklung in Deutschland: Die Zukunft Dauerhaft Umweltgerecht Gestalten*; Erich Schmidt Verlag: Berlin, Germany, 2002. (In German)

37. Network of Heads of European Environment Protection Agencies. *Delivering the Sustainable Use of Natural Resources*; Network of Heads of European Environment Protection Agencies: Copenhagen, Denmark, 2006.

38. European Environment Agency. *Well-Being and the Environment: Building a Resource-Efficient and Circular Economy in Europe*; European Environment Agency: Copenhagen, Denmark, 2014; pp. 1–52.

39. Schumacher, E. *Small Is Beautiful: A Study of Economics as if People Mattered*; Blond & Briggs: London, UK, 1973.

40. Kristof, K.; Kanthak, J.; Müller, F.; Rechenberg, B. What matters 2012: Resource efficiency—A key skill for sustainable societies. In *What Matters—Annual Report of the German Environment Agency*; German Environment Agency (UBA): Dessau-Roßlau, Germany, 2012; pp. 34–57.

41. Organization for Economic Cooperation and Development. *Material Resources, Productivity and the Environment*; Organization for Economic Cooperation and Development: Paris, France, 2015; pp. 1–176.

42. European Commission. *Communication from the Commission to the Council and the European Parliament—Towards a Thematic Strategy on the Sustainable Use of Natural Resources*; (COM(2003) 572 Final); European Commission: Brussels, Belgium, 2003.

43. Association of German Engineers (VDI). *Vdi 4800 Part 1: Resource Efficiency—Methodological Principles and Strategies*; Beuth Verlag GmbH: Berlin, Germany, 2016; pp. 1–54.

44. Kosmol, J.; Kanthak, J.; Herrmann, F.; Golde, M.; Alsleben, C.; Penn-Bressel, G.; Schmitz, S.; Gromke, U. *Glossar zum Ressourcenschutz*; Umweltbundesamt: Dessau-Roßlau, Germany, 2012; pp. 1–44. (In German)

45. Organization for Economic Cooperation and Development. *Measuring Material Flows and Resource Productivity. The Oecd Guide*; The Organisation for Economic Co-operation and Development (OECD): Paris, France, 2007.

46. Fischer-Kowalski, M.; von Weizsäcker, E.U.; Ren, Y.; Moriguchi, Y.; Crane, W.; Krausmann, F.; Eisenmenger, N.; Giljum, S.; Hennicke, P.; Kemp, R.; et al. *Decoupling Natural Resource Use and Environmental Impacts from Economic Growth*; United Nations/International Resource Panel: Nairobi, Kenya, 2011; pp. 1–174.

47. European Statistical System. *Economy-Wide Material Flow Accounts and Derived Indicators—A Methodological Guide*; Office for Official Publications of the European Communities: Luxembourg, 2001; p. 85.

48. Balzer, F.; Bünger, D.B.; Dauert, U.; Drosihn, D.; Eckermann, D.F.; Gellrich, A.; Geupel, M.; Gniffke, P.; Günther, J.; Hintzsche, M.; et al. *Data on the Environment 2015—Environmental Trends in Germany*; Federal Environment Agency (UBA): Dessau-Roßlau, Germany, 2015; pp. 1–144.

49. Buyny, S.; Lauber, U. *Further Development of the Indicator "Raw Material Productivity" in the National Strategy for Sustainable Development. Calculating Imports and Exports in Raw Material Equivalents*; Federal Statistical Office: Wiesbaden, Germany, 2010.

50. Federal Statistical Office of Germany. *Environmental-Economic Accounting, Sustainable Development in Germany, Environmental and Economic Indicators*; Federal Statistical Office: Wiesbaden, Germany, 2014.

51. Schoer, K.; Wood, R.; Arto, I.; Weinzettel, J. Estimating raw material equivalents on a macro-level: Comparison of multi-regional input-output analysis and hybrid lci-io. *Environ. Sci. Technol.* **2013**, *47*, 14282–14289. [CrossRef] [PubMed]

52. Schoer, K.; Giegrich, J.; Kovanda, J.; Lauwigi, C.; Liebich, A.; Buyny, S.; Matthias, J. *Conversion of European Product Flows into Raw Material Equivalents*; Ishares Europe Developed Real Estate ETF: Heidelberg, Germany, 2012; pp. 1–148.

53. Eurostat. Material Flow Accounts—Flows in Raw Material Equivalents. Available online: http://ec.europa.eu/eurostat/statistics-explained/index.php/Material_flow_accounts_-_flows_in_raw_material_equivalents (accessed on 13 January 2017).

54. Mancini, L.; Benini, L.; Sala, S. Resource footprint of europe: Complementarity of material flow analysis and life cycle assessment for policy supp.ort. *Environ. Sci. Policy* **2015**, *54*, 367–376. [CrossRef]

55. Lettenmeier, M.; Liedtke, C.; Rohn, H. Eight tons of material footprint—Suggestion for a resource cap for household consumption in Finland. *Resources* **2014**, *3*, 488–515. [CrossRef]

56. Daly, H.E. *Toward a Steady-State Economy*; Herman, E.D., Ed.; W.H. Freeman & Co.: San Francisco, CA, USA, 1973.

57. Ayres, R.U. Industrial metabolism. In *Technology and Environment*; Ausubel, J., Sladovich, H.E., Eds.; National Academies Press: Washington, DC, USA, 1989.

58. Von Weizsäcker, E.U.; Lovins, A.B.; Lovins, L.H. *Factor Four: Doubling Wealth, Halving Resource Use: The New Report to the Club of Rome*; Earthscan Publications Ltd.: London, UK, 1997.

59. Schmidt-Bleek, F.; Klüting, R. *Wieviel Umwelt Braucht der Mensch?: MIPS—Das Maß für Ökologisches Wirtschaften/How Much Environment does Man Need?: MIPS—The Measure of Ecological Management*; Birkhäuser: Berlin, Germany, 1994 (In German).

60. Von Weizsacker, E.U.; Hargroves, K.; Smith, M.; Desha, C.; Stasinopoulos, P. *Factor Five: Transforming the Global Economy through 80% Improvements in Resource Productivity*; Earthscan Publications Ltd.: London, UK; Droemer Munich, Germany, 2009.

61. Spangenberg, J.H.; Hinterberger, F.; Moll, S.; Schütz, H. Material flow analysis, TMR and the MIPS concept: A contribution to the development of indicators for measuring changes in consumption and production patterns. *Int. J. Sustain. Dev.* **1999**, *2*, 491–505. [CrossRef]

62. Ritthoff, M.; Rohn, H.; Liedtke, C. *Calculating Mips: Resource Productivity of Products and Services*; Wuppertal Institute for Climate, Environment and Energy: Wuppertal, Germany, 2002.

63. Saurat, M.; Ritthoff, M. Calculating mips 2.0. *Resources* **2013**, *2*, 581–607. [CrossRef]

64. Factor 10 Club. *Carnoules Declaration*; Factor 10 Club: Carnoules, France, 1994.

65. Hinterberger, F.; Luks, F.; Schmidt-Bleek, F. Material flows vs. 'natural capital': What makes an economy sustainable? *Ecol. Econ.* **1997**, *23*, 1–14. [CrossRef]

66. Jeske, U.; Kopfmüller, J.; Wintzer, D. Book Review: Wieviel Umwelt Braucht der Mensch?: MIPS—Das Maß für Ökologisches Wirschaften; TATuP—Zeitschrift des ITAS zur Technikfolgenabschätzung. *TATuP* **1995**, *4*, 11–16. (In German). Available online: http://www.tatup-journal.de/tadn951_jeua95a.php (accessed on 30 November 2017).

67. Schmidt, M. Book Review: Wieviel Umwelt Braucht der Mensch?: MIPS—Das maß für Ökologisches Wirtschaften. Spektrum der Wissenscahft. *Spektrum der Wissenschaft* **1995**, *6*, 116. Available online: http://www.spektrum.de/magazin/wieviel-umwelt-braucht-der-mensch-mips-das-mass-fuer-oekologisches-wirtschaften/822379 (accessed on 30 November 2017). (In German)

68. Coenen, R.; Klein-Vielhauer, S.; Kopfmüller, J. Stellungnahme zur Studie des Wuppertal-Instituts für Klima, Umwelt und Energie: "Zukunftsfähiges Deutschland". Ein Beitrag zu einer global nachhaltigen Entwicklung. *TATuP Zeitschrift des ITAS zur Technikfolgenabschätzung* **1996**, *5*, 1. (In German)

69. Schmidt, M. Zu den Schutzzielen der Ressourceneffizienz. *uwf UmweltWirtschaftsForum* **2014**, *22*, 147–152. (In German) [CrossRef]
70. *Global Guidance for Life Cycle Impact Assessment Indicators*; UNEP/SETAC Life Cycle Initiative: Nairobi, Kenya, 2016; Volume 1.
71. Scheringer, M.; Hofstetter, P. Auswahl, Begründung und Vergleich von Schutzgütern in der Ökobilanz-methodik—Ein offenes Problem. In *Schutzgüter und Ihre Abwägung aus der Sicht Verschiedener Disziplinen*; Umweltnatur-und Umweltsozialwissenschaften ETH: Zürich, Switzerland, 1997; pp. 4–10. (In German)
72. Hofstetter, P.; Scheringer, M.; Hirsch, G.; Jäggi, C.; Kytzia, S.; Leimbacher, J.; Schaber, P.; Scheringer, M.; Schütz, J.; Seidl, I. *Schutzgüter und Ihre Abwägung aus der Sicht Verschiedener Disziplinen*; Eidgenössische Technische Hochschule (ETH): Zürich, Switzerland, 1997; pp. 1–48. (In German)
73. Consoli, F.; Allen, D.; Boustead, I.; Fava, J.; Franklin, W.; Jensen, A.; Oude, N.D.; Parrish, R.; Perriman, R.; Postlethewaite, D.; et al. *Guidelines for Life-Cycle Assessment: A "Code of Practice"*; Society of Environmental Toxicology and Chemistry (SETAC): Brussels, Belgium, 1993; p. 79.
74. Vadenbo, C.; Rørbech, J.; Haupt, M.; Frischknecht, R. Abiotic resources: New impact assessment approaches in view of resource efficiency and resource criticality—55th discussion forum on life cycle assessment, Zurich, Switzerland, April 11, 2014. *Int. J. Life Cycle Assess.* **2014**, *19*, 1686–1692. [CrossRef]
75. Drielsma, J.; Allington, R.; Brady, T.; Guinée, J.; Hammarstrom, J.; Hummen, T.; Russell-Vaccari, A.; Schneider, L.; Sonnemann, G.; Weihed, P. Abiotic raw-materials in life cycle impact assessments: An emerging consensus across disciplines. *Resources* **2016**, *5*, 12. [CrossRef]
76. Hauschild, M.Z.; Goedkoop, M.; Guinée, J.; Heijungs, R.; Huijbregts, M.; Jolliet, O.; Margni, M.; De Schryver, A.; Humbert, S.; Laurent, A.; et al. Identifying best existing practice for characterization modeling in life cycle impact assessment. *Int. J. Life Cycle Assess.* **2013**, *18*, 683–697. [CrossRef]
77. Steen, B. *A Systematic Approach to Environmental Priority Strategies in Product Development (Eps) Version 2000—Models and Data of the Default Method*; Chalmers University of Technology, Centre for Environmental Assessment of Products and material Systems (CPM): Gothenburg, Sweden, 1999.
78. Dewulf, J.; Benini, L.; Mancini, L.; Sala, S.; Blengini, G.A.; Ardente, F.; Recchioni, M.; Maes, J.; Pant, R.; Pennington, D. Rethinking the area of protection "natural resources" in life cycle assessment. *Environ. Sci. Technol.* **2015**, *49*, 5310–5317. [CrossRef] [PubMed]
79. Hofstetter, P. *Perspectives in Life Cycle Impact Assessment: A Structured Approach to Combine Models of the Technosphere, Ecosphere and Valuesphere*; Kluwer Academic Publishers: Amsterdam, The Netherlands, 1998.
80. European Commission, Joint Research Centre. *International Reference Life Cycle Data System (ILCD) Handbook: Framework and Requirements for Life Cycle Impact Assessment Models and Indicators*; Publications Office of the European Union: Luxembourg, 2010.
81. Jolliet, O.; Müller-Wenk, R.; Bare, J.; Brent, A.; Goedkoop, M.; Heijungs, R.; Itsubo, N.; Peña, C.; Pennington, D.; Potting, J.; et al. The lcia midpoint-damage framework of the unep/setac life cycle initiative. *Int. J. Life Cycle Assess.* **2004**, *9*, 394–404. [CrossRef]
82. European Commission, Joint Research Centre. *International Reference Life Cycle Data System (ILCD) Handbook: Recommendations for Life Cycle Impact Assessment in the European Context—Based on Existing Environmental Impact Assessment Models and Factors*; Publications Office of the European Union: Luxembourg, 2010.
83. Thompson, M.; Ellis, R.; Wildavsky, A.B.; Wildavsky, M. *Cultural Theory*; Westview Press: Boulder, CO, USA, 1990.
84. Hellweg, S.; Hofstetter, T.B.; Hungerbühler, K. Discounting and the environment: Should current impacts be weighted differently than impacts harming future generations? *Int. J. Life Cycle Assess.* **2003**, *8*, 8–18.
85. Grünig, M.; Srebotnjak, T.; Schock, M.; Porsch, L.; Möller-Gulland, J. *Plakative und Schnelle Umweltinformation Mittels Hochaggregierter Kenngrößen zur Nachhaltigen Entwicklung*; Umweltbundesamt: Berlin, Germany, 2011.
86. Baccini, P.; Brunner, P.H. *Metabolism of the Anthroposphere: Analysis, Evaluation, Design*; The MIT Press: Cambridge, MA, USA, 2012; p. 408.
87. Ayres, R.U.; Ayres, L. *A Handbook of Industrial Ecology*; Edward Elgar Publishing: Cheltenham, UK, 2002; p. 688.
88. Ayres, R.U. Industrial metabolism and global change. *Int. Soc. Sci. J.* **1989**, *41*, 363–373.
89. Bringezu, S.; Sand, I.V.D.; Schütz, H.; Bleischwitz, R.; Moll, S. *Analysing Global Resource Use of National and Regional Economies Across Various Levels*; Bringezu, S., Ed.; Greenleaf: Sheffield, UK, 2009; pp. 10–51.

90. Bundesregierung. *Nationale Nachhaltigkeitsstrategie—Fortschrittsbericht 2012*; Bundesregierung: Berlin, Germany, 2012; pp. 1–264.

91. Smeets, E.; Weterings, R. *Environmental Indicators: Typology and Overview*; European Environment Agency (EEA): Copenhagen, Denmark, 1999.

92. Hertwich, E.; van der Voet, E.; Suh, S.; Tukker, A.; Huijbregts, M.; Kazmierczyk, P.; Lenzen, M.; McNeely, J.; Moriguchi, Y. *Assessing the Environmental Impacts of Consumption and Production: Priority Products and Materials, a Report of the Working Group on the Environmental Impacts of Products and Materials to the International Panel for Sustainable Resource Management*; UNEP: Paris, France, 2010.

93. Association of German Engineers (VDI). *VDI 4800 Part 2: Resource Efficiency—Evaluation of the Use of Raw Materials*; Beuth Verlag GmbH: Berlin, Germany, 2016; pp. 1–38.

94. German Environment Agency (UBA). ProBas-Process-Oriented Basic Data for Environmental Management Systems. Available online: http://www.probas.umweltbundesamt.de/ (accessed on 27 October 2017).

95. Wernet, G.; Bauer, C.; Steubing, B.; Reinhard, J.; Moreno-Ruiz, E.; Weidema, B. The ecoinvent database version 3 (part I): Overview and methodology. *Int. J. Life Cycle Assess.* **2016**, *21*, 1218–1230. [CrossRef]

96. Srebotnjak, T.; Polzin, C.; Giljum, S.; Herbert, S.; Lutter, S. *Establishing Environmental Sustainability Thresholds and Indicators*; Ecologic Institute and SERI: Berlin, Germany, 2010; pp. 1–139.

97. Giegrich, J.; Liebich, A.; Lauwigi, C.; Reinhardt, J. *Indicators for the Use of Raw Materials in the Context of Sustainable Development in Germany*; 01/2012; German Environment Agency (UBA): Dessau-Roßlau, Germany, 2012; pp. 1–248.

98. Berger, M.; Finkbeiner, M. Correlation analysis of life cycle impact assessment indicators measuring resource use. *Int. J. Life Cycle Assess.* **2010**, *16*, 74–81. [CrossRef]

99. Voet, E.; Oers, L.; Nikolic, I. Dematerialization: Not just a matter of weight. *J. Ind. Ecol.* **2004**, *8*, 121–137. [CrossRef]

100. Huijbregts, M.A.J.; Rombouts, L.J.A.; Hellweg, S.; Frischknecht, R.; Hendriks, A.J.; van de Meent, D.; Ragas, A.M.J.; Reijnders, L.; Struijs, J. Is cumulative fossil energy demand a useful indicator for the environmental performance of products? *Environ. Sci. Technol.* **2006**, *40*, 641–648. [CrossRef] [PubMed]

101. Huijbregts, M.A.J.; Hellweg, S.; Frischknecht, R.; Hendriks, H.W.M.; Hungerbühler, K.; Hendriks, A.J. Cumulative energy demand as predictor for the environmental burden of commodity production. *Environ. Sci. Technol.* **2010**, *44*, 2189–2196. [CrossRef] [PubMed]

102. Hertwich, E.G.; Van der Voet, E.; Tukker, B.; Tukker, A. *Assessing the Environmental Impacts of Consumption and Production*; United Nations Environment Programme: Paris, France, 2010; pp. 1–108.

103. German Environment Agency (UBA). Domestic Extraction and Imports. Available online: https://www.umweltbundesamt.de/daten/rohstoffe-als-ressource/inlaendische-entnahme-von-rohstoffen#textpart-1 (accessed on 17 January 2017).

104. Steinmann, Z.J.N.; Schipper, A.M.; Hauck, M.; Huijbregts, M.A.J. How many environmental impact indicators are needed in the evaluation of product life cycles? *Environ. Sci. Technol.* **2016**, *50*, 1–23. [CrossRef] [PubMed]

105. Drielsma, J.A.; Russell-Vaccari, A.J.; Drnek, T.; Brady, T.; Weihed, P.; Mistry, M.; Simbor, L.P. Mineral resources in life cycle impact assessment—Defining the path forward. *Int. J. Life Cycle Assess.* **2016**, *21*, 85–105. [CrossRef]

106. Hotelling, H. The economics of exhaustible resources. *J. Political Econ.* **1931**, *39*, 137–175. [CrossRef]

107. Costanza, R.; Cumberland, J.; Daly, H.; Goodland, R.; Norgaard, R. *An Introduction to Ecological Economics*; St. Lucie Press: Boca Rato, Nepal, 1997.

108. Riedel, E.; Janiak, C. *Anorganische Chemie*; De Gruyter: Berlin, Germany, 2011.

109. Diederen, A. *Global Resource Depletion—Managed Austerity and the Elements of Hope*; Eburon Academic Publishers: Delft, The Netherlands, 2010.

110. Skinner, B.J. A second iron age ahead? The distribution of chemical elements in the earth's crust sets natural limits to man's supply of metals that are much more important to the future of society than limits on energy. *Am. Sci.* **1976**, *64*, 258–269.

111. Skinner, B.J. Exploring the Resource Base. Presented at the Workshop on "The Long-Run Availability of Minerals", Washington, DC, USA, 22–23 April 2001; pp. 1–25.

112. Bardi, U.; Randers, J. *Extracted: How the Quest for Mineral Wealth is Plundering the Planet*; A Reportnto the Club of Rome/Ugo Bardi; White River Junction: Windsor County, VT, USA; VT Chelsea Green Publishing: River Junction, VT, USA, 2014; p. 29.

113. Giurco, D.; Prior, T.; Mudd, G.; Mason, L.; Behrisch, J. *Peak Minerals in Australia: A Review of Changing Impacts and Benefits*; Institute for Sustainable Futures University of Technology, Sydney & Department of Civil Engineering (Monash University): Sydney, Australia, 2010; pp. 1–110.

114. Northey, S.; Mudd, G.M.; Weng, Z.; Mohr, S.; Giurco, D. Modelling future copp.er ore grade decline based on a detailed assessment of copp.er resources and mining. *Resour. Conserv. Recycl.* **2014**, *83*, 190–201. [CrossRef]

115. Hubbert, M.K. Nuclear energy and the fossil fuels. In *Drilling and Production Practice*; American Petroleum Institute: New York, NY, USA, 1956; pp. 1–40.

116. Sorrell, S.; Speirs, J.; Bentley, R.; Brandt, A.; Miller, R. *Global Oil Depletion: An Assessment of the Evidence for Near-Term Peak in Global Oil Production*; UK Energy Research Centre: London, UK, 2009; pp. 1–228.

117. Sverdrup, H.U.; Ragnarsdottir, K.V.; Koca, D. An assessment of metal supp.ly sustainability as an input to policy: Security of supply, extraction rates, stocks-in-use, recycling, and risk of scarcity. *J. Clean. Prod.* **2017**, *140*, 359–372. [CrossRef]

118. May, D.; Prior, T.; Cordell, D.; Giurco, D. Peak minerals: Theoretical foundations and practical application. *Nat. Resour. Res.* **2012**, *21*, 43–60. [CrossRef]

119. Giraud, P.-N. A note on hubbert's thesis on mineral commodities production peaks and derived forecasting techniques. *Procedia Eng.* **2012**, *46*, 22–26. [CrossRef]

120. Rustad, J.R. Peak nothing: Recent trends in mineral resource production. *Environ. Sci Technol.* **2012**, *46*, 1903–1906. [CrossRef] [PubMed]

121. Graedel, T.; Gunn, G.; Tercero Espinoza, L. Metal resources, use and criticality. In *Critical Metals Handbook*; John Wiley & Sons: Hoboken, NJ, USA, 2014; pp. 1–19.

122. Wellmer, F.W. Reserves and resources of the geosphere, terms so often misunderstood. Is the life index of reserves of natural resources a guide to the future? *Z. Dtsch. Ges. Geowiss.* **2008**, *159*, 575–590.

123. Marscheider-Weidemann, F.; Langkau, S.; Hummen, T.; Erdmann, L.; Espinoza, L.T.; Angerer, G. *Rohstoffe für Zukunftstechnologien 2016—Raw Materials for Emerging Technologies 2016*; German Mineral Resources Agency DERA at the Federal Institute for Geosciences and Natural Resources BGR: Berlin, Germany, 2016; pp. 1–360.

124. Schaffartzik, A.; Mayer, A.; Eisenmenger, N.; Krausmann, F. Global patterns of metal extractivism, 1950–2010: Providing the bones for the industrial society's skeleton. *Ecol. Econ.* **2016**, *122*, 101–110. [CrossRef]

125. U.S. Geological Survey. *Mineral Commodity Summaries 2016*; U.S. Geological Survey: Reston, VA, USA, 2016; pp. 1–205.

126. Graedel, T.E.; Harper, E.M.; Nassar, N.T.; Nuss, P.; Reck, B.K. Criticality of metals and metalloids. *Proc. Natl. Acad. Sci. USA* **2015**, *112*, 4257–4262. [CrossRef] [PubMed]

127. Melcher, F.; Wilken, H. Die Verfügbarkeit von Hochtechnologie-Rohstoffen. *Chem. Unserer Zeit* **2013**, *47*, 32–49. [CrossRef]

128. Gutzmer, J.; Klossek, A. Die Versorgung mit wirtschaftskritischen Rohstoffen—Eine Ursachensuche und -Analyse. In *Strategische Rohstoffe—Risikovorsorge*; Kausch, P., Bertau, M., Gutzmer, J., Matschullat, J., Eds.; Springer: Berlin, Germany, 2014; pp. 61–73.

129. Nassar, N.; Graedel, T.; Harper, E. By-product metals are technologically essential but have problematic supp.ly. *Sci. Adv.* **2015**, *1*, 1–10. [CrossRef] [PubMed]

130. European Commission. *Report on Critical Raw Materials for the EU—Critical Raw Material Profiles*; European Commission: Brussels, Belgium, 2014; pp. 1–205.

131. Graedel, T.E.; Barr, R.; Chandler, C.; Chase, T.; Choi, J.; Christoffersen, L.; Friedlander, E.; Henly, C.; Jun, C.; Nassar, N.T.; et al. Methodology of metal criticality determination. *Environ. Sci. Technol.* **2011**, *46*, 1063–1070. [CrossRef] [PubMed]

132. Buijs, B.; Sievers, H. *Critical Thinking About Critical Minerals—Assessing Risks Related to Resource Scarcity*; Federal Institute for Geosciences and Natural Resources BGR: Hannover, Germany, 2011; pp. 1–19.

133. Buijs, B.; Sievers, H.; Espinoza, L.A.T. Limits to the critical raw materials app.roach. *Proc. ICE Waste Res. Manag.* **2012**, *165*, 201–208.

134. Gandenberger, C.; Glöser, S.; Marscheider-Weidemann, F.; Ostertag, K.; Walz, R. *Die Versorgung der Deutschen Wirtschaft Mit Roh—und Werkstoffen für Hochtechnologien—Präzisierung und Weiterentwicklung der Deutschen Rohstoffstrategie*; Arbeitsbericht Nr. 150; TAB—Büro für Technikfolgen-Abschätzung beim Deutschen Bundestag: Berlin, Germany, 2012; pp. 1–238. (In German)

135. Erdmann, L.; Graedel, T.E. Criticality of non-fuel minerals: A review of major approaches and analyses. *Environ. Sci. Technol.* **2011**, *45*, 7620–7630. [CrossRef] [PubMed]

136. European Commission. *Report on Critical Raw Materials for the Eu*; European Commission: Brussels, Belgium, 2014; pp. 1–41.

137. Angerer, G.; Buchholz, P.; Gutzmer, J.; Hagelüken, C.; Herzig, P.; Littke, R.; Thauer, R.K.; Wellmer, F.-W. *Rohstoffe für die Energieversorgung der Zukunft: Geologie—Märkte—Umwelteinflüsse*; Deutsche Akademie der Technikwissenschaften (acatech): Berlin, Germany, 2016.

138. Angerer, G.; Marscheider-Weidemann, F.; Lüllmann, A.; Erdmann, L.; Scharp, M.; Handke, V.; Marwede, M. *Raw Materials for Emerging Technologies*; The Influence of Sector-Specific Feedstock Demand on Future Raw Materials Consumption in Material-Intensive Emerging Technologies. Final Report Abridged; Fraunhofer Institute for Systems and Innovation Research ISI: Berlin, Germany, 2009; pp. 1–19.

139. Rüttinger, L.; Sharma, V. *Cimate Change and Mining*; Adelphi Research: Berlin, Germany, 2016.

140. Rüttinger, L.; Scholl, C.; Sharma, V.; Vogt, R.; Kämper, C. Klimress: Impacts of climate change on the ecological criticality of germany's raw material demand (fkz: 3716 48 324 0). In *Environmental Research Plan (UFOPLAN) of the Federal Ministry for the Environment, Nature Conservation, Building and Nuclear Safety*; German Environment Agency, Ed.; Adelphi Research & Ifeu—Institute for Energy and Environmental Research & The University of Queensland—Centre for Social Responsibility in Mining (CSRM): Dessau-Roßlau, Germany, 2017.

141. Federal Ministry of Education and Research (BMBF). *Raw Materials of Strategic Economic Importance for High-Tech Made in Germany—BMBF Research and Development Programme for New Raw Material Technologies*; Federal Ministry of Education and Research (BMBF): Bonn, Germany, 2013; pp. 1–64.

142. Coulomb, R.; Dietz, S.; Godunova, M.; Nielsen, T.B. *Critical Minerals Today and in 2030*; OECD Publishing: Paris, France, 2015.

143. Gemechu, E.D.; Sonnemann, G.; Helbig, C.; Thorenz, A.; Tuma, A. Import-based indicator for the geopolitical supply risk of raw materials in life cycle sustainability assessments. *J. Ind. Ecol.* **2016**, *20*, 154–165. [CrossRef]

144. Tuma, A.; Reller, A.; Thorenz, C.; Kolotzek, C.; Helbig, C. *Nachhaltige Ressourcenstrategien in Unternehmen: Identifikation Kritischer Rohstoffe und Erarbeitung Von Handlungsempfehlunge zur Umsetzung Einer Ressourceneffizienten Produktion*; University of Augsburg: Augsburg, Germany, 2014; pp. 1–63.

145. Dehoust, G.; Manhart, A.; Vogt, R.; Kämper, C.; Giegrich, J.; Auberger, A.; Priester, M.; Dolega, P. *Discussion of the Environmental Limits of Primary Raw Material Extraction and Development of a Method for Assessing the Environmental Availability of Raw Materials to Further Develop the Criticality Concept (ÖkoRess I)—Summary*; German Environment Agency (UBA): Dessau-Roßlau, Germany, 2017; pp. 1–31.

146. Dehoust, G.; Manhart, A.; Vogt, R.; Kämper, C.; Giegrich, J.; Priester, M.; Rechlin, A.; Rüttinger, L.; Scholl, C. ÖkoRess II: Further development of policy options for an ecological raw materials policy. In *Environmental Research Plan (UFOPLAN) of the Federal Ministry for the Environment, Nature Conservation, Building and Nuclear Safety*; German Environment Agency, Ed.; Öko Institut—Institute for Applied Ecology; Ifeu—Institute for Energy and Environmental Research; Projekt-Consult; Adelphi Research: Dessau-Roßlau, Germany, 2017.

147. Rüttinger, L.; Scholl, C. *Responsible Mining? Challenges, Perspectives and Approaches—Summary of the Findings of the Research Project "Approaches to Reducing Negative Environmental and Social Impacts in the Production of Metal Raw Materials"*; German Environment Agency (UBA): Dessau-Roßlau, Germany, 2017; pp. 1–37.

148. Glöser, S.; Tercero Espinoza, L.; Gandenberger, C.; Faulstich, M. Raw material criticality in the context of classical risk assessment. *Resour. Policy* **2015**, *44*, 35–46. [CrossRef]

149. Behrendt, S.; Scharp, M.; Erdmann, L.; Kahlenborn, W.; Feil, M.; Dereje, C.; Bleischwitz, R.; Delzeit, R. *Rare Metal: Measures and Concepts for the Solution of the Problem of Conflict-Aggravating Raw Material Extraction—the Example of Coltan*; German Environment Agency (UBA): Dessau-Roßlau, Germany, 2007.

150. Franken, G.; Vasters, J.; Dorner, U.; Schütte, P.; Küster, D.; Nüher, U. Certified Trading Chains in Mineral Production. In *Competition and Conflicts on Resource Use. Natural Resource Management and Policy*; Hartard, S., Liebert, W., Eds.; Springer International Publishing: Cham, Switzerland, 2015; Volume 46, pp. 177–186.

151. Dupuy, R.; Roman, P.; Mougenot, B. Analyzing socio-environmental conflicts with a commonsian transactional framework: Application to a mining conflict in peru. *J. Econ. Issues* **2015**, *49*, 895–921. [CrossRef]

152. Tänzler, D.; Westerkamp, M. *Rohstoffkonflikte Nachhaltig Vermeiden*; German Environment Agency (UBA): Dessau-Roßlau, Germany, 2010.

153. Schandl, H.; Fischer-Kowalski, M.; West, J.; Giljum, S.; Dittrich, M.; Eisenmenger, N.; Geschke, A.; Lieber, M.; Wieland, H.P.; Schaffartzik, A.; et al. *Global Material Flows and Resource Productivity*; An Assessment Study of the UNEP; International Resource Panel; United Nations Environment Programme: Paris, France, 2016; p. 200.

154. BIO by Deloitte. *Study on Data for a Raw Material System Analysis: Roadmap and Test of the Fully Operational Msa for Raw Materials*; Prepared for the European Commission, DG GROW; BIO by Deloitte: Neuilly-sur-Seine, France, 2015; pp. 1–179.

155. United Nations. *Rio Declaration on Environment and Development*; United Nations General Assembly: Rio de Janeiro, Brazil, 1992.

156. Bringezu, S.; Schütz, H.; Moll, S. Rationale for and interpretation of economy-wide materials flow analysis and derived indicators. *J. Ind. Ecol.* **2003**, *7*, 43–64. [CrossRef]

157. European Commission. *Communication from the Commission on the Precautionary Principle*; COM(2000) 1 Final; European Commission: Brussels, Belgium, 2000; pp. 1–28.

158. Rockström, J.; Steffen, W.; Noone, K.; Persson, Å.; Chapin, S.I.; Lambin, E.; Lenton, T.M.; Scheffer, M.; Folke, C.; Schellnhuber, H.J. Planetary boundaries: Exploring the safe operating space for humanity. *Ecol. Soc.* **2009**, *14*, 1–33. [CrossRef]

159. Peduzzi, P. Sand, rarer than one thinks. In *UNEP Global Environmental Alert Service (GEAS)*; United Nations Environment Programme: Nairobi, Kenya, 2014; pp. 1–15.

160. European Commission. Stakeholder Consultation—Options for Resource Efficiency Indicators. Available online: http://ec.europa.eu/environment/resource_efficiency/targets_indicators/stakeholder_consultation/index_en.htm (accessed on 13 January 2017).

161. Federal Office for the Environment. *Development of Switzerland's Worldwide Environmental Impact*; Federal Office for the Environment: Bern, Switzerland, 2014.

MDPI

St. Alban-Anlage 66

4052 Basel

Switzerland

Tel. +41 61 683 77 34

Fax +41 61 302 89 18

www.mdpi.com

Resources Editorial Office

E-mail: resources@mdpi.com

www.mdpi.com/journal/resources

www.ingramcontent.com/pod-product-compliance
Lightning Source LLC
Chambersburg PA
CBHW051900210326
41597CB00033B/5966